GOLD MINING IN WASHINGTON STATE

- YESTERDAY AND TODAY -

"Pioneer Stories"

Hiram "Okanogan" Smith
The Ballard Brothers

AUTHOR

Ken Martin in Collaboration with Vida Martin

Golden Treasures Publishing

Published by
Golden Treasures Publishing
P.O. Box 1703
Stanwood, WA 98292

Editor: Veronica Freund

ISBN: 0-9643300-0-8

Library of Congress Catalog Card Number:
94-78727

Commercial Printing, Wenatchee, Washington

FRONT COVER

Ross Lake now covers much of the upper Skagit drainage area which was heavily mined in the late 1800's. Prior to the opening of the horse trail up the Skagit River in 1892, miners traveled down this corridor from Hope, B.C. to reach the placer diggings on Ruby Creek. The scenic view is looking north from the North Cascades Highway 20. Mount Hozomeen, near the Canadian border, can be seen in the distance.

Bottom Left: Chain-bucket dredge which operated on Swauk Creek near Liberty, Washington, in the 1920's. The dredge was the largest of its kind in Washington State. Tailings from the workings can still be seen scattered along Swauk Creek which parallels Highway 97. Over $35,000 was recovered by the dredge from 1926 to 1929. For further information about the development of the chain-bucket (bucket-line) dredge, see page 101. (John Juleen photo).

Bottom Center: The gold pan contains several large nuggets found in the Northwest. The pendant nugget, in the center of the pan, was found on Ruby Creek at the turn of the century.

Bottom Right: A young miner poses with a drifter drill in the United Copper Mine located on the western slope of Eagle Mountain, about five miles Northeast of Chewelah, Washington, 1914. Ore from the mine contained gold, silver and copper. With gold at $20.67 a fine ounce, silver at 53¢ a fine ounce, and copper at 13.2¢ a pound, the average gross value of the ore shipped during 1908 was $9.596 a ton. Notice the candle used as a source of light. Felt hats were sometimes worn by workers in the mines. The hat was warm but provided little protection from jagged or falling rock. (Plate VII, U.S.G.S. Bulletin, No. 500).

BACK COVER

The color-penciled map of the Swauk and Peshastin (Blewett) districts shown on the back cover was found among a collection of old documents in Seattle, Washington. The map corresponds closely with map details featured in the Atlas of the United States, Mount Stuart Folio, Washington; U.S.G.S., 1904. It may have been used as a field map. The X's near Liberty probably indicate placer and hardrock mines. (The Johanson collection).

ACKNOWLEDGMENTS

The author wishes to acknowledge all of those individuals, living and not living, who have over the years contributed to the information and pictures included in this book. These include miners, wives and children of miners, historians, geologists, museum and library personnel, and numerous authors. The list of those who have contributed to the research has grown so large that it would be difficult to mention all of them by name without missing someone. I would, however, like to give special recognition to my wife, Vida, who assisted in research and interviews and editing of the text. The book would not have been possible without her effort.

The numerous friendships generated from interviews and the various pursuits of information are in themselves worthy of the whole endeavor. I will always remember and prize these friendships.

THE AUTHOR AND COAUTHOR

I was raised as a coal miner's son in Boone, Iowa. Over the past twenty years, as time would allow, I have enjoyed the challenges and rewards of recreational prospecting, mostly in eastern Washington. Two summers were spent with family and friends in Alaska prospecting for gold on the South Fork of the Forty Mile River.

At the time of the publishing of this book, Vida, the coauthor, is an elementary school librarian. She completed the M.A. in Curriculum and Instruction at Western Washington University. I completed the B.A. and M.A. at Seattle Pacific University.

I dedicate this book to the memory of my father, George "Doc" Martin, who worked thirty-three years as an Iowa coal miner. He was a strong but gentle man, the last to mine coal in a family descended from Welch coal miners. Dad lived by the rules of honesty and hard work. I will always remember his blackened fingernails, not black from coal dust, but rather from the bruises incurred while working underground. Having been raised in a miner's home, I have learned to appreciate the toil and hardship shared by all of those who have earned their living from mining, whether it be coal, gold, or minerals of another kind.

AUTHORS, KEN AND VIDA. The submersible dredge was purchased around 1962 from Mel's (treasure hunter Mel Fisher's) Aqua Shop in Redondo Beach, California. With a removable sluice tray, this unique underwater separator was considered state of the art during the beginning years of recreational dredging. Vida is holding a large sixteen-inch gold pan passed down from the 1800's. It was typical for that era. An experienced prospector could process 100 pans of this size in a ten-hour day. Our German Shepherd, Duke, enjoys being a part of things.

TRIBUTE

Stories surrounding the lure of gold and adventures in finding it comprise an important chapter from the annals of early Washington State history. A few remaining old-timers still gather around tables in small-town coffee shops to reminisce about the good old days or the time when granddad came west to discover gold and a better way of life. Only a generation removed, these gentlemen talk with familiarity about nineteenth century gold strikes at places such as Liberty and Blewett, Rich Bar on the Similkameen, and Ruby Creek in the North Cascades. Today, a number of these historic locations are identified by placards which tell sketches of information about a time when the country was still largely unsettled.

In museums throughout the state, simplistic hardware from gold rush days is being preserved for the children of future generations. Such artifacts serve as object lessons of a time when ingenuity, unending courage, and hard work were the major ingredients of survival. Behind the places with unusual names and tools of unusual shapes and sizes, there once lived real men and women committed to making the best of a difficult way of life. For their contribution to our pioneer heritage, we owe a debt of appreciation.

As part of this tribute, I would like to quote a portion of an article published in the Everett News under the title, "What Mining Has Done, " January 22, 1892:

In 1849, in our country the Missouri River was the border land of civilization; 1500 miles of trackless deserts and unexplored mountain ranges, infested with wild beasts . . . lay between the western verge of civilization and the new El Dorado on the Pacific coast. But no such barrier could stay the tide of hardy adventurers that swarmed across the continent in great caravans of brave, determined men and heroic women. Through infinite labor and privations they reached the promised land. The tide of emigration spread all over the mountain ranges and valleys of the West, and in 40 years has produced more than $2,000,000,000 in gold and silver, actual money, created and added to our country's resources. But this grand output of the precious metals is not the only wonderful result obtained. It enabled our people to explore and settle a vast country, more than half as large as Europe; it has given eight new states to the Union and three grand territories, unrivaled in their resources of mineral wealth and variety of soil and climate by any other country of like size on the globe. These grand and beneficent results of precious metal mining stand without a parallel in human history.

"TO BOOM OR NOT TO BOOM"

To boom or not to boom - this was the question posed in an editorial published in the Snohomish weekly, The Eye, in May of 1884.

In their effort to present the latest news from the mining districts, early newspapers sometimes unknowingly became the vehicles of exaggerated rumors. Prospectors were eager to publish their mining claims in the hopes of being able to sell them. Merchants owning stores in the path of the migration were delighted to have a "boom." It meant sales and profits. So they, too, were eager to pass on any hopeful rumors. Newspapers increased their sales many-fold when they headlined the news of new and rich gold strikes. There were occasions, though, when they had to retract statements which were later proven to be in error. Some of these supposed rich finds were subsequently considered a "humbug."

Editors of The Eye were approached to "boom" the Sultan after new discoveries had been made there in 1884. The stories were second and third hand so the newspaper refused to publish the news until a reporter had been sent to gather the real facts. The editors had remembered previous times when reports had been exaggerated so were compelled to be cautious. Such was the case in 1880 when the Skagit discoveries were blown out of proportion.

Concerning the news information gathered for this book, the reports so tediously researched for day, month, and year may not always represent the facts as they actually happened. They do, however, represent the excitement that prevailed as miners did make important discoveries in new places.

TABLE OF CONTENTS

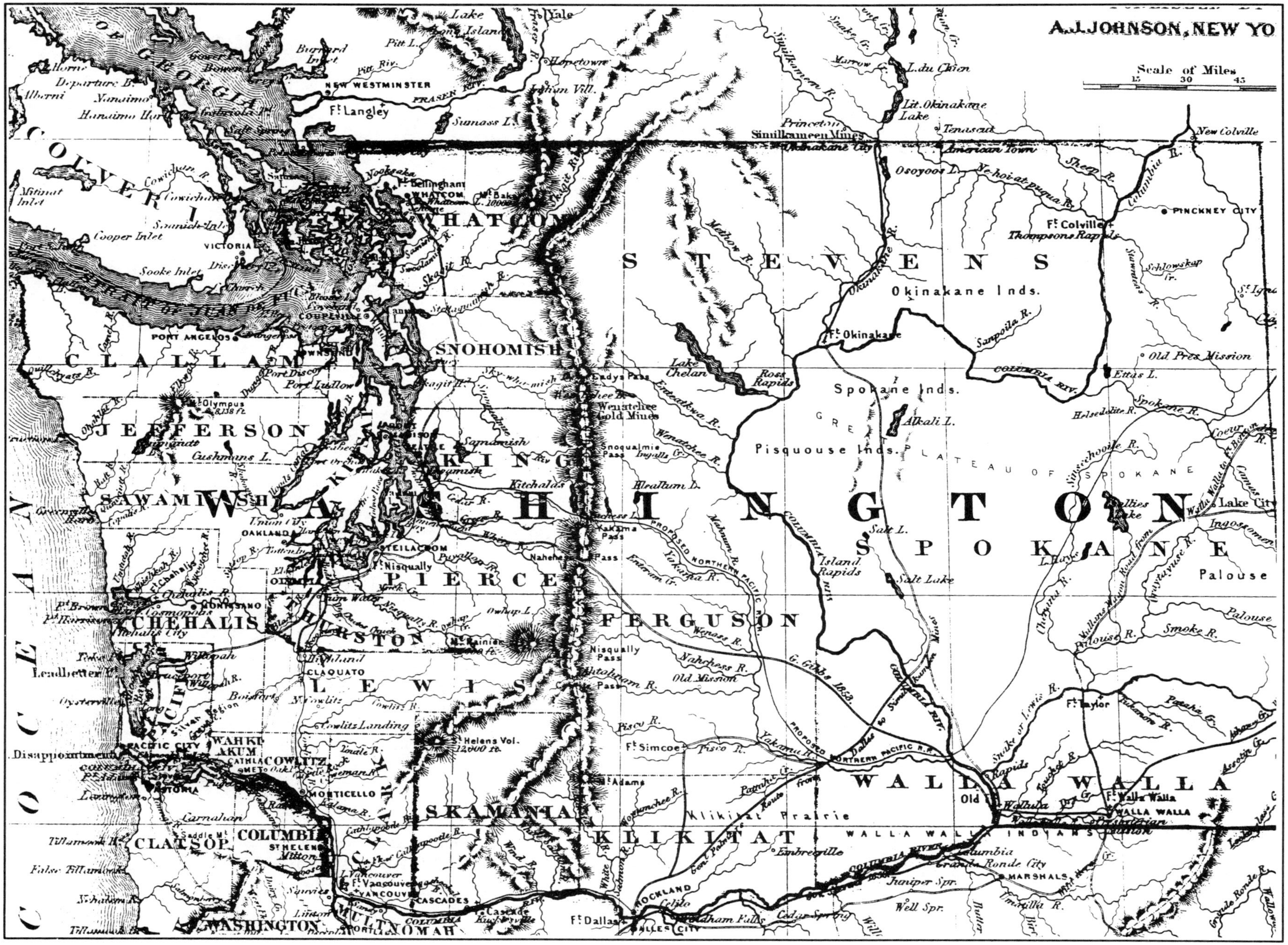

Washington Territory section of Johnson's Oregon and Washington Map, 1865, showing early settlements, travel routes, and forts.

PART I

THE LURE OF GOLD

Over the years, prospectors have faced the most hostile and demanding of conditions in their search for gold. The stories of starvation and hardship as told by those returning from the Yukon gold rush should have been sufficient to discourage any newcomer. Nevertheless, they continued into the North Country by the thousands at the end of the 19th century. The possibility of hitting pay dirt, however remote, was all the lure the stampeders needed. In a time when so many knew only the bare necessities of life, just the thought of striking it rich provided an imaginary reprieve from the harsh realities of life. Each new gold rush gained momentum as tales of new-found riches circulated between camps and filtered through to the outside.

Long before Washington was a state, adventurous prospectors were exploring uncharted northwest wilderness country in search of the lustrous yellow metal. For these rugged individualists, there was no such thing as the end of the trail. The elusive pot at the end of the rainbow always lay just beyond the next ridge, or around the next bend, or in a steep canyon where no one else dared to go. Among those who persisted in the toilsome search, a fortunate few were rewarded with fabulous discoveries in yet unnamed places.

For most, the pot of gold slowly took shape over a period of weeks and months as the raw metal was tediously extracted from the ground. As reality would have it, Gold - beautiful, glittering gold did not fall magically from the sky to be scooped up by some opportunistic dreamer. It was the precious result of long days of heavy labor in unfriendly, sordid conditions. Following hopeful signs, shadowy figures maneuvered about in the muck and grime of dark, dank tunnels to move tons of rock and debris. By these means, an occasional prized specimen emerged into the daylight to capture the delight of wide-eyed observers. Thus, expectations and rumors were kept alive.

The various challenges of the nineteenth century prospector were aptly described under the heading "Prospecting for Gold," The Eye, December 14, 1893:

> *Every spring hundreds of prospectors plunge into this wilderness on a search for gold and silver, some on their own account, but most of them go grub-staked by local capitalists, that is, all their expenses are paid and in return they give a half interest in whatever mines they may discover. With a rifle, a hatchet, a bag of flour and a frying pan slung over his shoulder, the prospector starts on his quest. He goes as far as possible by stage or carriage, then on foot enters the forest and follows for many miles a narrow trail that winds in and out between the trunks of great trees, whose heads, reaching up hundreds of feet, shut off the sun and keep the earth on a twilight so deep that only moss and fungi can grow. At times the path skirts the side of a rushing mountain river, where in every eddy are dozens of darting trout. Then it climbs and winds high up the mountain side, grows fainter, and finally disappears, leaving the gold hunter to scramble on over fallen logs and around precipices as best he can.*
>
> *When night comes he stops usually in the bottom of a valley by a stream of melted snow water, fries and eats some pancakes, cuts a great pile of fir boughs for a bed, throws himself on it and is asleep in a moment. All through the night the cold and damp from the stream and the mossy ground creep up and in the morning he is stiff and scarcely able to stagger on. He continues thus day after day, seeing only the trees near about him, and occasionally a bit of dark mountain side through a rift in their tops.*

Finally he has climbed high up and comes out above the forests, and sees again for the first time in days the full daylight. Here, high above the world on the mountain sides, he stands (hikes) weeks searching for silver and gold veins and hunting goat and deer for meat. Sometimes he loses himself, his food gives out, and he starves, or by a mis-step falls thousands of feet from a precipice and is dashed to pieces in the valley below. If fortunate, he locates several claims, stakes them out, finds bearings for them on the peaks above, and then, loaded with specimens makes his way back to civilization where his ores are assayed. He has brought the best specimens only, so they come out splendidly. People become excited, and a party, including mine speculators, goes up to see what he has found. If good, the mines are sold on the spot to speculators, who in turn sell to rich eastern or English mining companies. These cut trails through the woods, send up machinery piecemeal on donkey or mule back, hire a lot of miners at three dollars a day, and the mine is started for profit or loss, according to the way the lead holds out.

THE FIRST DISCOVERY OF GOLD

In October of 1853, Captain George B. McClellan met with the governor of the Washington Territory at Fort Colville. McClellan had been assigned the task of finding a suitable railroad route through the Cascade Mountains. During the meeting, McClellan told Governor Stevens that in September of that year traces of gold had been found in the Yakima Valley. Stevens recorded from the information provided by McClellan:

On the third of September the main camp was moved to the Yakima valley, where another depot was made, while Captain McClellan, with Mr. Gibbs, Minter, and six men went to examine the passes at the head of the river and Lieutenant Duncan was left in charge to examine as much as possible of the neighboring country. Here the first traces of gold were discovered, and though not sufficiently abundant to pay for working, it caused considerable excitement in the camp. A large band of Indians, under Ouhi, a brother of Kamiakan, encamped near, and were very friendly, this chief being the most good natured Indian yet seen in the country. [1]

In his report, Stevens later mentioned other discoveries made in the region:

It is also worthy of observation that gold was found to exist, in the explorations of 1853, throughout the whole region between the Cascades and the main Columbia to north of the boundary, and paying localities have since been found at several points, particularly on the Southern tributary of the Wenatshapam (Wenatchee). The gold quartz also is found on the Nachess River, the gold bearing crossing the Columbia and stretching eastward along Clark's Fork and the Koutenay river, unquestionably extends to the Rocky Mountains. [2]

According to the July 31, 1858, American Railroad Journal, news spread of the discoveries which had been made in the region explored by Captain McClellan. In 1854, the general public had become aware of the gold deposits which existed in the North Cascades:

In the following year (1854), much additional information was gained in reference to the gold in that territory. The marvelous Indian stories in regard to rich diggings had as much relation to the Cascade country of

Washington as to that north of the 49th parallel. In 1854, the public mind became impressed with the belief that it required only a diligent prospecting of the country to develop the richness of its gold deposits. [3]

McClellan's finds in 1853 were probably the first non-Indian discoveries of gold in the region. By the mid-1850's, there was considerable activity along the tributaries of the upper Columbia in the vicinity of Fort Colville. The discovery of gold on the Pend d'Oreille was reported to have been by a half-breed named Walker. The initial reports coming from that area told of good success. In the summer of 1855, miners near the mouth of the Pend d'Oreille were making from $2 to $10 a day. Six miles up the river on a bar, a company of miners worked six to eight days and brought out $4 to $19 to the hand. In 1914, the scholar, William J. Trimble, wrote:

> *. . . a movement occurred to the vicinity of Ft. Colville, which had some of the characteristics of a genuine miner's "rush," and which ushered in the gold era in the inland Empire. . . . Considerable numbers of the citizens of Oregon and Washington participated in this movement and prospected in the Colville mines in the fall of 1855.*[4]

During July and August of 1855, over 1000 miners made their way to the Fort Colville vicinity, some from the Puget Sound over the Cascades and others by the line of the Columbia. Most retraced their steps because of the fear of hostile Indians, but the few who remained made good wages during the following two years.

In 1857, prospectors explored further north and discovered gold along the Fraser and Thompson Rivers. The Hudson's Bay Company had interior posts in the upper Fraser district and in spite of Indian hostilities, kept the trail open north from Fort Colville. Gold was also discovered in the Fort Hope and Fort Yale vicinity of the Fraser. During August of 1858, an estimated 3000 miners were at work on the river near Fort Hope. Six men hit a paystreak and took out $600 in six hours using only a rocker. Above Fort Hope, on Texas Bar, two sluices were being worked on the same bar averaging not over $5 a day to the man. On Hill's Bar, the largest before the canyon, miners with rockers were averaging 1½ ounces per man per day. The fifteen or twenty rockers being cleaned up were each showing from two to three ounces of gold.[5]

In 1859, gold was discovered on the Similkameen, a tributary of the upper Okanogan River. During the 60's, more discoveries were made in a number of stream locations east of the Cascade summit, including Swauk and Peshastin Creeks. In August of 1878, eight white men and twelve Chinese were working twelve miles up from the mouth of Swauk Creek. A good wagon road ran from Ellensburg to the mines. A prospector by the name of Humbolt had been averaging $3 per day and had found two large nuggets, one weighing $89 and the other $13.50. Moses Bolman, with three helpers, recovered $238 in one day. The largest nugget found that summer was by a Mr. Bates. It was valued at $189.[6]

As of July, 1858, on the west side of the summit, discoveries had been made at the headwaters of the Skagit, Snoqualmie, Skokomish, and Duwamish rivers. [7] There is evidence that prospectors were on the Sultan in the early 1860's and that the river was heavily mined during the 1870's. During the summer of 1870, about 250 miners were on the river. At that time, an old rocker and rocker iron were found in such a condition that they fell apart when moved. It was believed that the rocker had sat idle since around 1861. In the spring of 1884, there were about fifty Chinese working on the river in small crews. It was estimated that each were averaging from 75¢ to $2.00 per day. Two prospectors, Dr. Moudy and Wm. Beach, while prospecting on the Horseshoe claim, carried dirt in a bucket from the hill to the creek and in one hour washed out $1.50 with a pan. Around 1890, a miner by the name of E. G. Payne found an $8.00 nugget in the ancient river channel of the Sultan. Given the price of gold at that time, it probably weighed about ½ ounce.

As of 1878, an estimated $10,000 to $20,000 had been mined from the Sultan River. In the late 1890's, the Sultan was still considered one of the best producers in the region:

> *. . . Sultan River, a branch of the Skykomish, is the only stream wherein systematic placer mining has been carried on for any length of time. Some work was done there thirty years ago. The greatest amount of gold was taken out from fifteen to twenty years ago . . . No one knows how much gold has been found on the Sultan. It is supposed that the amount taken therefrom is greater than that derived from any other Puget Sound stream . . . It would be safe to say that over $100,000 in gold dust has come from these mines. Possibly two or three times that sum has been derived therefrom.* [8]

Some of the discoveries in the North Cascades were made as miners moved south from the gold fields of British Columbia. Thousands of miners from California had stampeded into Canada after a shipment of Fraser River gold arrived in San Francisco in February of 1858. By this time, much of the surface gold in California had already been worked out so miners were eager to move on to something better. Also, there was the theory circulating that prospects improved the further north one went. Reports coming out of the Fraser River told of rich finds so little encouragement was needed for miners to drop everything and head north.

After each new discovery, news traveled by word-of-mouth. The stories which circulated from camp to camp were no doubt embellished with spurious details of huge nuggets and instant riches. Hopeful of finding better pay dirt, miners frequently abandoned profitable diggings to search elsewhere. This was the case along the Similkameen River in northcentral Washington Territory. Men of the Boundary Commission surveying the International Line discovered gold along the river in 1859. Hundreds of prospectors flocked to the banks of the Similkameen. By 1860, a settlement of tent and shanty dwellings, called Okanogan City, with several hundred residents had sprung up near the river. When news was received of new discoveries in Canada, the miners packed up their belongings to follow the new leads. Almost overnight, the boomtown was abandoned. After a few months, though, many returned to harvest the rich values of the Similkameen.

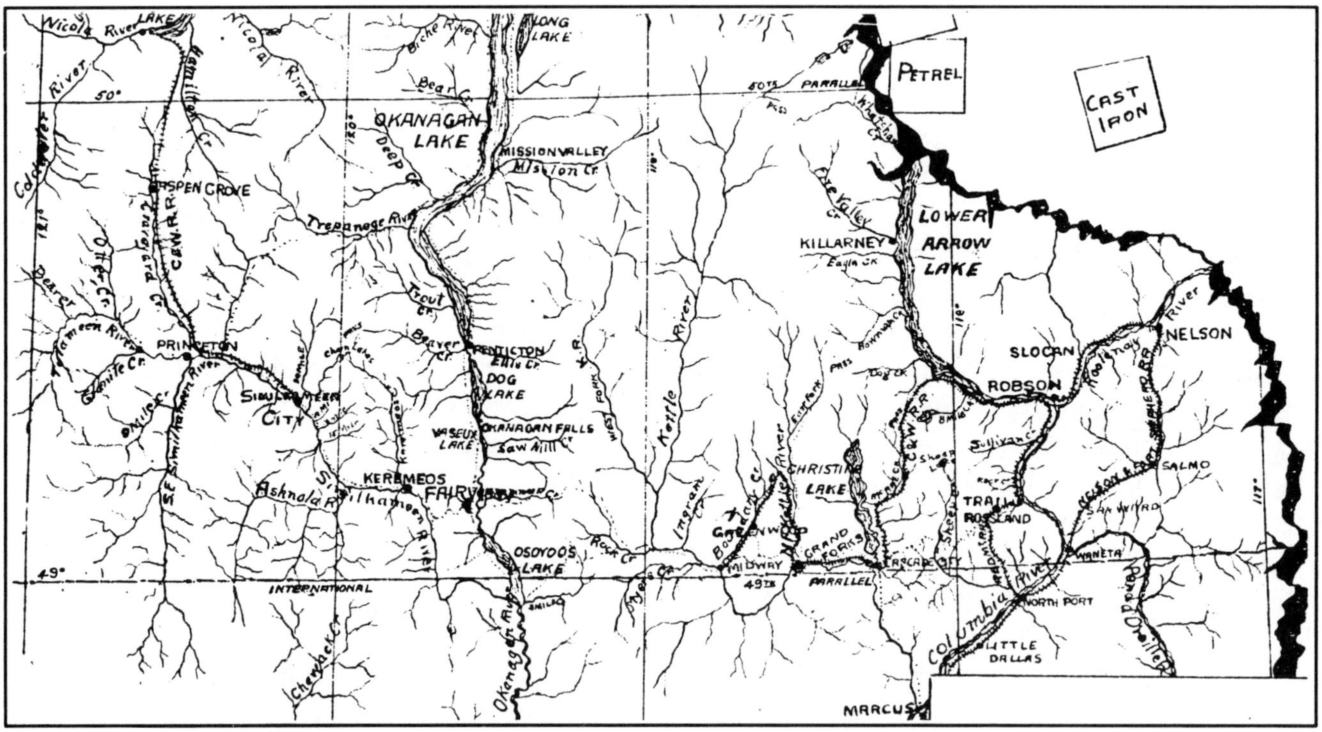

Mining map showing the Similkameen River, Osoyoos Mining Division, British Columbia. Compiled by Frank Bailey, 1900.

MINERS AND NATIVE AMERICANS CLASH

In the early days, there were many perils to be considered by those who dared to journey into gold country. Unpredictable weather systems presented life-threatening challenges in the form of impassable rivers, deep snow, and severe temperatures. While passing over Chilkoot Pass into the Klondike, more than sixty stampeders perished when caught in a massive snowslide. In some locales, Native Americans took exception to the intrusion of the white man. This was the case in northcentral Washington Territory when a band of Indians attacked a mining party enroute from Walla Walla to the Fraser River in Canada.

The rush of miners into the interior of Washington Territory was met with opposition from various Indian tribes:

> *Discovery of gold in the vicinity of Fort Colville incited a stampede to that country. This was in the spring of 1855. At that period, Governor Stevens was making his famous eastern tour through the territory engaged in treaties and agreements with the various tribes, and this gold discovery so excited the members of his escort that it was with difficulty they were prevented from deserting. . . . The first note of defiance was sounded by Pierre Jerome, chief of the Kettle Falls Indians, about August 1, 1855. He decided emphatically that no white man should pass through his country, this declaration was soon followed by the Yakimas.*[9]

In an effort to bring the Yakimas, Walla Wallas, Nez Perces and other eastern tribes into treaty agreements, Governor Stevens summoned a great council to meet at Walla Walla in the summer of 1855. General Joel Palmer, Superintendent of Indian Affairs for Oregon Territory, joined Governor Stevens in the effort. About 5000 interior Indians followed their chiefs, sub-chiefs, and delegates to the Walla Walla Valley where they set up their tepees. After considerable deliberation, treaties were signed designating three reservations and large annuitics to be paid to the tribes. Widespread suspicion and regret over the treaties by the Indians is believed to have been a major factor in the war of 1855-56. An incident which fueled hostility was the attack on a party of prospectors traveling through the Yakima Valley enroute to the Colville mines. Two of the party were shot. Three others escaped by hiding during the day and traveling at night. When the three reached Seattle, they reported the news of their tragedy. The fate of another party which had previously disappeared was now made apparent.

General Wool, commander of the Department of the Pacific, believed the treaties of 1855 were a major cause of the war. He was especially opposed to the Governor's volunteer militia who were in his opinion out to plunder the Indians. To address those whom he felt had no right on Indian lands, he issued an order in which some allowance was made for gold miners attempting to reach the Colville mines:

> *He issued the following order: "No emigrants or other whites, except the Hudson's Bay Company, or persons having ceded rights from the Indians, will be permitted to settle or remain in the Indian country, or on land not ceded by treaty, confirmed by the Senate and approved by the President of the United States. . . . These orders are not, however, to apply to the miners engaged in collecting gold at the Colville mines. The miners will, however, be notified that should they interfere with the Indians, or their squaws, they will be punished or sent out of the country."* [10]

BATTLE AT McLOUGHLIN CANYON

McLoughlin Canyon, located about four miles south of Tonasket, Washington, east of the Okanogan River.

The attack at McLoughlin Canyon in the summer of 1858 was one of numerous isolated incidents in which local bands of Indians exhibited hostility toward whites traveling through their native territories. David McLoughlin had arranged to lead a caravan of miners, traders, and packers north to the Fraser River country. David was the son of the renowned Dr. John McLoughlin who had served as the Columbia District head for the Hudson's Bay Company. The story of the battle and events leading to it were recalled by McLoughlin in an interview published in the Spokane Review, May 24, 1891:

> *It came near being a massacre. We started from Wallula in the later part of June, 1858, with a pack train and 149 men for the Fraser River country. The outfit comprised a representation from nearly all the states, and quite a number of half-breeds. We got along peaceably with an occasional quarrel among the different sets of our men (which, of course, didn't count), until we arrived at Moses Canyon, where we were attacked by Red Jacket, chief of the Palouse Indians. In the fight we lost one man killed (Evans, of Portland), and several wounded, besides some of our pack animals. The reason we got off so lightly was that the Indians were anxious to stampede the stock, especially the pack animals, instead of hunting scalps. We knew for certain we were in for it for the rest of the journey. . . .*

Francis Wolff, a merchant and outfitter from the Fort Colville area, was traveling with the McLoughlin party. Years later, in 1904, Wolff was asked to write a sketch of his life history. In his recollections, he included details of the ambush which add to the story as told by McLoughlin. He remembered that they left Walla Walla on about the the 20th of July. When the pack train came to the Columbia opposite from Fort Okanogan, they were warned by Francois Duchouquette, who was in charge of the fort, and Indian Chief Pierre that they should not continue north. The Indians were prepared to fight if they persisted in their journey. After lengthy discussions which lasted through the night, canoes were provided and the entire brigade crossed the river. According to McLoughlin, Chief Moses had taken command of the combined Indian forces and was opposed to their crossing the river.

While going from the river to the Fort, an Indian was noticed wearing the clothes and riding a horse belonging to one of their party. The man had not been missed prior to the miners reaching the river. It was supposed that he had tried to return to Walla Walla. The victim and others of the party had been warned that the country was not safe and that they should stay close to the rest of the party.

When the company of riders left the fort, an advance and rear guard of twenty-five men each were appointed. The pack train traveled north along the banks of the Okanogan River to the point where they would have to either cross to the west side or take the more direct route - a mile long canyon a short distance from the river. A steep bluff ran from the canyon to the river's edge forcing pack trains to either cross the river at this point or take the canyon and then return to the river a few miles upstream. McLoughlin described the approach to the canyon and the battle that followed:

... the next morning after we had climbed the first hill, before entering the canyon, and not seeing any signs of Indians, I became suspicious and called a halt, while I rode forward with one man. I had not proceeded 200 yards when I noticed bushes piled against rocks, and my eyes being pretty sharp, I noticed that the leaves were wilted. Telling my companion to stop where he was I started to investigate the suspicious circumstance, and had got within thirty yards when I noticed a painted buck behind a little stone fort and breastplate, and before I could investigate any further or bring my gun to my shoulder, he fired, the ball taking effect in the neck of my horse, killing him instantly. The fight immediately became general and lasted from 10 o'clock a.m., until 5 p.m., when we retreated to the river under a steady fire from Indians ... An attempt was made several times during the night to stampede our stock, but as each attempt cost the reds some of their best braves, they desisted toward morning. We lost in the fight four killed and twenty wounded ... also, twenty-five pack animals ... The next morning we built a raft of dead driftwood and crossed to the west side of the river. That is what they call the massacre of McLoughlin's canyon ... and it was hot for a while. We were followed by the Indians all the way to Rock Creek, occasionally getting a crack at some thoughtless straggler.

The Pioneer and Democrat printed a brief account of the battle:

We learn through Messrs. Kent and Smith's Express that on August 4th and 5th an engagement took place between 180 whites, under the command of Mr. David McLoughlin of Oregon, and 80 Indians, on the forks of the Okinikan and Similkameen rivers. Three whites were killed and two wounded mortally, while two were slightly hurt. Not more than 25 whites were engaged in the fight at any one time. Six Indians were wounded. The cause of the difficulty was a robbery of some cattle by the Indians from Mr. Wolf of Colville Valley. [11]

A humorous sequel to the story of the McLoughlin battle has been passed down in the historical accounts as told by some of the local Indians. As the story goes, the Indians liberated from the spoils of the white man a sack of peculiar looking beans. At their camp, they expectantly waited as the beans cooked to produce what was hoped to be a savory dish. The beans boiled and boiled. When more than ample time had elapsed, instead of a nourishing dish of palatable bean soup, the pot yielded a distasteful brew of very strong coffee. The final comment was - "and they couldn't eat it."

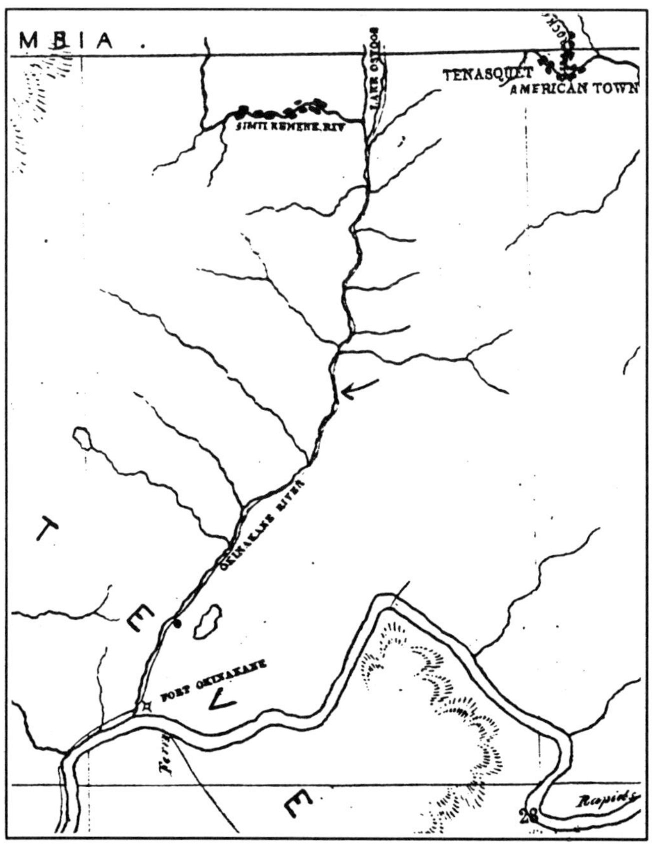

Portion of Map of Public Survey in Washington Territory, 1870. McLoughlin Canyon was located about four miles below present-day Tonasket, Washington, on the Cariboo Trail which led to the gold fields of the upper Fraser River. (Arrow indicates approximate location of the canyon)

Hazard Ballard's horses crossing the cable suspension bridge which spanned the Skagit River just above the mouth of Ruby Creek, 1935. The horse trail continued east along Ruby Creek to the high country bordering the Cascade Divide. The numerous mines of the Slate Creek District could be reached by traveling this trail. The place where the bridge crossed the river is now under the backwater of Ross Dam. INSET: Hazard (Passenger side) and his brother, Leon. (Photos courtesy of Bill Wehmeyer)

GOLD IS DISCOVERED ON RUBY CREEK

Scenic wonders of unparalleled beauty awaited those who dared to trek through the untamed wilderness. Majestic mountains with snowcapped peaks towered above what must have appeared as an endless maze of rivers and valleys. Compelled by the lure of gold and adventure, determined prospectors overcame obstacles of immense proportions to reach their destinations.

In the early 1860's, miners exploring south from Canada discovered gold on Ruby Creek, a tributary of the upper Skagit. Ruby Creek is located in some of the most difficult country in the North Cascades. Getting into the district presented a problem since there was no horse trail up the Skagit River. For years, miners and merchants talked of the need for such a trail. During the spring of 1892, miners were still traveling the British Columbia route to reach Ruby Creek. A trail ran south from Fort Hope where miners outfitted with pack horses, camp supplies, and miner's utensils.

The horse bridge at the "Devil's Corner," also called the "Devil's Elbow," on the Skagit River, 1921. Prior to the completion of a horse trail in 1892, the steep rock cliffs in this section of the river was a barrier to those traveling up the gorge. It was called the Devil's Elbow because, at this location, the river made a sharp turn almost at a right angle. The infamous name was used in the 1870's and possibly earlier. (Darius Kinsey photo from Johanson collection)

By the end of July, 1892, a horse trail with bridges had been built from Goodell's landing on the lower Skagit to Ruby Creek. Mining supplies were brought by canoe as far as Goodell's, then carried by pack horse the rest of the distance. As soon as the trail was open for travel, hydraulic equipment was brought in to work the steep placer hillsides. One of the flumes which channeled water to work claims near the mouth of Canyon Creek was a mile long. Another flume and ditch prior to construction by the Ruby Hydraulic Gold Company was anticipated to be three miles long when finished.

In 1892, the Everett News featured an article with the heading, "HIGH PLACER BARS - Miners washing out $50 a day in the Ruby Creek Diggings." As indicated by the article, miners had become alert to the fact that rich quartz veins were to be found in the surrounding hills:

> *Messrs. Collie and McMillian are engaged in actual placer mining near the mouth of Ruby Creek. They are working on false bed-rock near an old deposit. They have only primitive appliances, using no quicksilver and only two sluice boxes. They, with only the assistance of one man, cleaned up $1260 worth of gold in twenty-four days - an average of over fifty dollars a day. Meantime they have so protected their works as to make them good for another thousand. One nugget washed out of these diggings weighed over seven dollars. It is the opinion of conservative miners that the great placer deposits of Ruby Creek, covering a large territory, are surrounded by extensive quartz deposits, as rich as any to be found in the Northwest.* (12)

PART II
HARDROCK MINING

For many years, prospectors have referred to the Mother Lode as the source which they greatly aspire to discover. Such is the hope - to look up one day and see an outcropping of quartz shot through with native gold. Story books are filled with tales of prospectors who stumbled upon the richest gold deposit one can imagine, only to have it lost by one circumstance or another. Early prospectors vigorously dug wherever they thought a rich vein might be uncovered. The mountain slopes of Washington and other western states are pocketed with thousands of short tunnels which produced little. When traces of gold were found, in most cases they soon played out or were of such little value that it simply was not worth the effort. Interest was kept alive, though, by the occasional find of very rich ore.

RUBY - THE FIRST OKANOGAN COUNTY SEAT

In those districts where gold reserves seemed to be in great supply, temporary shelters soon gave place to more permanent dwellings. Small towns took shape around trading posts and near ear-shattering stamp mills. Though many of these pioneer towns have since disappeared, a few remnants of buildings and artifacts still remain as vestiges to a bygone era.

Ruby, Washington at its peak in March, 1891. Ruby was a mining boomtown in the early 1890's and the first Okanogan County seat. The town was located along Salmon Creek in a valley about thirteen miles northwest of the present-day town of Okanogan. (Photo courtesy of Ann Briley)

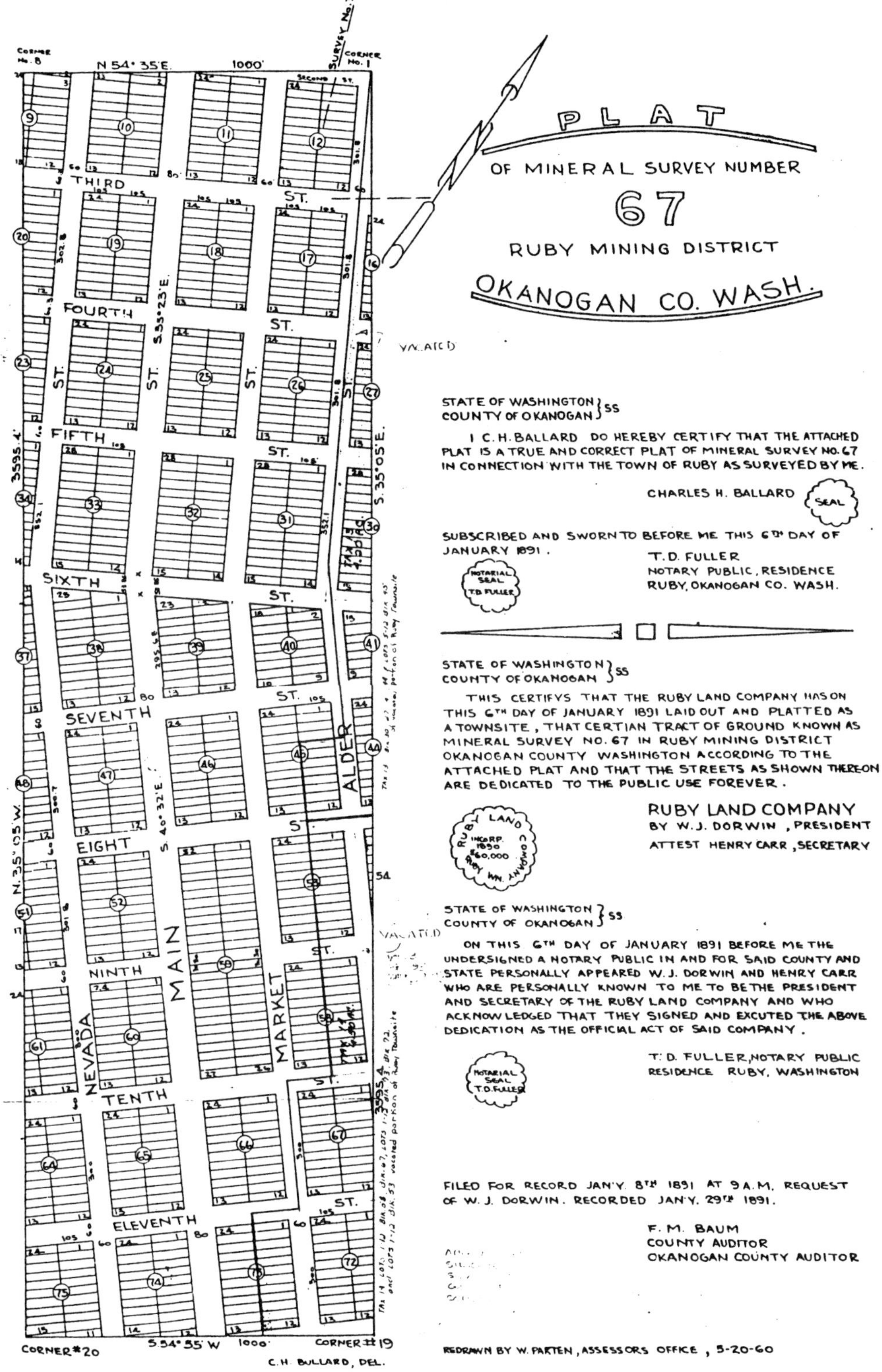

Mineral Survey Plat, number 67, of the town of Ruby in the Ruby mining district. U.S. Mineral Surveyor, Charles H. Ballard, attested to the correctness of the plat. The document was recorded on January 29, 1891.

At the beginning of the 1890's, Ruby was one of the most active mining camps in the state. Most of the buildings were of frame construction, utilizing dimensional rough-sawn lumber. This method of building was much faster and easier than using logs. The town bustled with energy - wagons going to and fro, children playing in back yards, and miners gathering at the community hall to talk about the latest prospects. Businesses of all kinds provided goods and services to a surrounding population estimated to be from 2,000 to 3,000 people. After Okanogan County was formed, Ruby became the temporary county seat from May of 1888, until February, 1889, when Conconully (formerly Salmon City) a few miles up the road became the seat of government.

As was typical of mining towns, Ruby was a rough place. Brothels, saloons, and gambling tables provided the setting for brawls and occasional murders, involving both whites and Indians. A local brewery supplied lager beer for the town's saloons, guessed to be anywhere from six to twenty. Main street boiled with dust as thirsty miners crowded into town after a hard day at the mines. Behind the dust and noise, though, there were consciences at work to create something good and permanent. As a move toward decency in 1892, the town council instructed the marshal to force all brothel businesses off the main street, possibly engendered by the shooting of a former patron by one of the madams.

The County Superintendent of Education, Mrs. Virginia Grainger, had opposed the building of a school at Ruby because of the small number of prospective students. In spite of the opposition, local citizens undertook to build a fine school building which also doubled as a community hall. Mrs. Grainger taught at the school for a year. (1)

Rumors of the local prospects had traveled all the way to the coast. The August 3, 1889, issue of the Snohomish weekly, The Eye, highlighted in its news column:

> *A big excitement prevails in the Conconully and Ruby districts over the discovery of a rich deposit of chloride ore near the former camp. The hills in the vicinity are said to be fairly alive with excited prospectors, and hundreds of locations are being made.*

Numerous silver-gold ore deposits were discovered in the hills above Salmon Creek in 1886, the year that the Moses Reservation was opened to mineral entry. A mining district was organized with the recording office at Ruby. Some of the initial assays ran very high and so ran the hopes of investors and miners. A concentrating mill was built one-half mile east of town to process ore delivered by a mile-long tramway from the First Thought mine. According to George A. Bethune, the first state geologist, assays from the surface of the Arlington mine ran as high as $1,500 to the ton. Having summarized samplings from several other mines, including the Ruby, First Thought, Fourth of July, War Eagle, and Idaho, geologist Bethune presented an encouraging word about the future of the district:

> *In Ruby district I noted a score or more of promising prospects, the result each of the indefatigable pursuit for mineral of those hardy prospectors who are doing so much toward advertising Ruby district's great promise as an ore producer to the world. All inspected by me impelled me to inform their locators, that in my opinion, the game is worth the candle.* (2)

This word of encouragement was prior to the depression of 1893 when the price of silver plummeted. Also, as was typical of a number of mines in the Okanogan, ores in the mines around Ruby became lower-grade at deeper levels. Processing the ore was difficult. Several years later, Arthur Homer Fischer in a thesis, published as Bulletin No. 4, presented a postmortem note of optimism as he described the complex ores of the area:

> *The treatment of low-grade ores of the Ruby and Conconully districts has been difficult, owing to their complex nature, but during the last two years laboratory metallurgical investigations have given such encouraging results as to make it seem probable that these ores may be profitably worked in the near future. . .* (3)

When the depression of 1893 came and the market for silver suffered, the town of Ruby lost its economic base and could no longer survive. The mines ceased operation, miners moved out, and local businesses shut down. Abandoned buildings were left to be plundered and dismantled for use at other locations. Some of the boards were taken to the Methow where a schoolhouse was built. Today, about all that marks the townsite, except for a historic information sign, is a few foundation remnants scattered across the landscape.

While Ruby went into a decline, Conconully continued to survive in spite of the local mining slump and other catastrophes which befell the community. A fire consumed the town in 1892 and a flash flood devastated its buildings in 1894. Even though it was a mining town, it was a desirable place to live. Newspaperman, James Hagerty, described the town as it was prior to the fire that engulfed it in 1892:

> *Conconully is the home of Lieutenant Governor Laughton and is one of the pleasantest spots in the mountains. It is different from most mining camps, as here miners and prospectors have determined to make their homes and have brought the gentle and refining influence of women and home to the camp which has had the effect of doing away with all lawlessness. No man can conceive a prettier place. Here the valley of Salmon Creek widens out nearly a mile in extent. To the west Mineral Hill rises gradually until the peaks are lost in the clouds; to the north and south the mountains rise abruptly, but are covered with timber which is of an evergreen verdure, while to the east lies Conconully Lake, a beautiful sheet of water 1,500 feet wide and three miles long, with gently sloping banks on either side.* (4)

Success or Failure: In the early 1890's, a Senate Committee on Mines and Mining issued a report on the cost of producing precious metals. The report suggested that a very high percentage of all supposed mining discoveries and locations had turned out to be a failure. Mining in general had not been a profitable industry:

> *It may be safely stated that if all the necessary expenditures of prospecting, developing, extracting and reducing gold and silver ore are taken into consideration, excluding foolish and wildcat speculations, mining for the precious metals is prosecuted at a greater average loss than is suffered by any other business in the United States, and that it would not be prosecuted at all if it were not for the occasional discoveries from which large fortunes are rapidly acquired.* (5)

From reading numerous writings on the subject, I have observed that most of the early lode mine failures were attributed to certain factors. One of these was the fault nature of ore bodies. Sometimes miners would simply come to the end of an ore vein which had been fractured and shifted elsewhere. Being unable to relocate the displaced vein, they abandoned the mine.

Profits at some of the mines were adversely affected by milling practices which did not yield a high recovery. Numerous mills were built to treat free-milling ore by amalgamation. By this method, only the richest gold ores could be worked profitably. Frequently, too, free-milling ore near the surface changed in character at deeper levels. Some mills added the use of cyanide to extract gold from non-amalgamable ores. In 1896,-a small cyanide plant was erected at the twenty-stamp mill at Blewett to recover gold from the tailings.

Another factor influencing mining was the smallness of ore bodies making commercial development infeasible. Also, mines were usually most productive near the surface where they were more easily worked and required less machinery and overhead expense. At deeper levels, the cost of production often exceeded the value of the metal produced. There were cases, too, where excessive amounts of water flooded the tunnels forcing miners to discontinue work.

There was also the problem of transportation of ore to processing facilities. Numerous mines were in isolated locations, sometimes miles from the nearest wagon road.

HIRAM "OKANOGAN" SMITH

Pioneer Prospector and Statesman

Geologist Bethune, 1891, credited Hiram Smith as having been the first to discover gold-bearing rock in the Washington Territory:

> *Among early pioneers in the gold hunting army that invaded Washington back in the sixties, was Hon. Hiram F. Smith, present representative in the legislature for Okanogan County. To Mr. Smith and a little band of prospectors is due the credit of having discovered the first gold bearing rock found in Washington. It was on the nation's natal day, July 4, 1871, that Mr. Smith and party camped at the base of Mt. Chopaka, a bold peak rising from an undulating surrounding country . . . .* [6]

Hiram "Okanogan" Smith

Considered by many to be the first pioneer settler in the Okanogan, Hiram Smith has been referred to as the "Father of the Okanogan." Smith was born in

1890 photo of Okanogan Smith holding his granddaughter, Ella. Notes on the back of the photo identify the following: Ella's father (Smith's son-in-law) John Evans standing in front of Smith. Scotch George, a prospector, standing next to John. George Runnels at the far right. Former Colville Indian agent, Major Gwydir is sitting on the porch. The gentleman at the left in front of the water barrel is not identified. (Photo courtesy of Bob Irwin)

Kennebec County, Maine, on June 11, 1829. He migrated with his parents to Iowa in 1837, to Illinois in 1841, then four years later went to Michigan where he learned the printer's trade at the Detroit Free Press. In 1848, he traveled to New York where he worked as a foreman of printers at the New York Tribune. This was during the time that Horace Greeley was with the Tribune.

Excited about the fantastic stories coming out of the 49' gold rush, Smith left his post at the Tribune to make the long journey to the gold fields of California. He worked his way across the Isthmus of Panama and up the coast to San Francisco. After nearly ten years in mining and other pursuits, he moved north to the Dalles in Oregon where he operated a freighting service. From the Dalles, he joined the migration of miners going north into British Columbia. After gold was discovered on the Similkameen in Washington Territory in 1859, he traveled to the Okanogan Valley where he settled at a location near Lake Osoyoos. While other dates have been suggested, evidence points to 1860 as the year he settled in the Okanogan. He had been at Rock Creek placer mining and running a small store. The gold rush to Rock Creek was going strong during the summer of 1860.

Smith prospected the hills of the lower Similkameen in search of gold and silver and posted claims at locations that were later to become active mining operations. The historian Steele wrote that Smith's first claim was in the early 70's on Mount Ellemeham, north of Loomis. It was called the "Julia," but later relocated as the "King Solomon." (7) Julia was the name of his daughter. In April of 1892, Smith applied for mining patents for the Caaba, Okanogan, Oswego, Summit, Eagle Quartz, State of the Maine, and Chopaka mines. He was known to have declared that there was enough silver in his Similkameen mines to pay off the national debt. The January 1, 1892, Spokane Review, told of the rather colorful method by which Smith brought public attention to his promising mines:

> *During the 70's, the Hon. H. F. Smith enlisted some persons in Olympia in a scheme to open some silver quartz mines lying north of the Similkameen river, near the boundary line, and some work was done in the way of development and 500 pounds of*

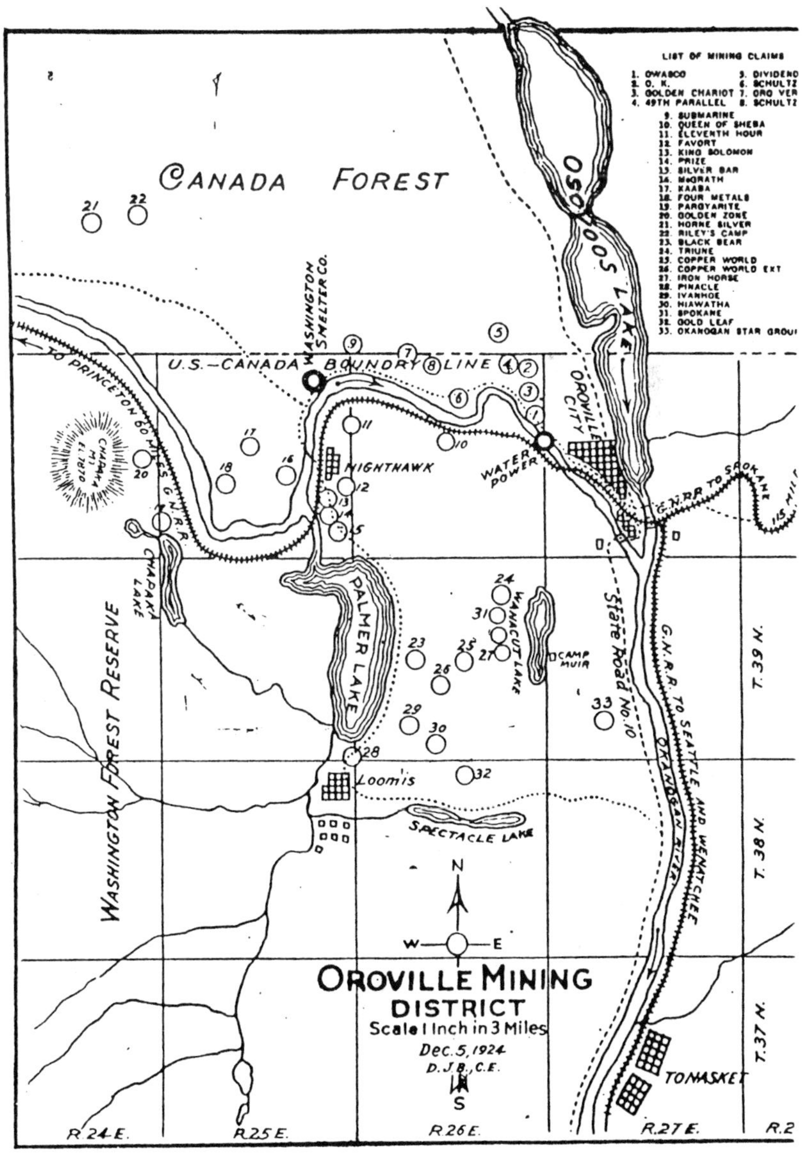

Oroville Mining District, encompassing the former Mount Chopaka-Similkameen Mining District.

ore was packed out and shipped to San Francisco where the silver was extracted and run into a bar, a portion of which was presented to President Hayes upon his visit to the territory in 1880. This, I believe, was the first silver bullion produced from the ores of Washington. [(8)]

One can only guess what President Hayes might have done with the sampling of Washington Territory silver. Could he have used it as a paperweight in his second floor office at the White House?

According to tribal custom, Smith married Mary Manuel, the daughter of a chief of the Okanogan Indians. He was regarded with high respect and was considered to be a trustworthy person among the Okanogan Indians. Chief Tonasket, of whom Smith had purchased 300 acres of land, once said of him, "When Okanogan Smith tells us to do anything he knows how to do it and shows us how; he never lies, while the government does." [(9)] Ann Briley, author and historian, has said of Smith:

> *He was a remarkable man . . . an educated and accomplished man who, without the spiritual calling of a missionary, nevertheless lived among the Indians straight and moral; moderate in all things; pleasantly spoken; an arbitor sought after where he might have been a rummy tyrant holding sway.* [(10)]

Smith was known to possess one of the finest ranches in the country. In the early 1860's he used pack horses to transport 1,200 fruit trees from Hope, B.C. to his ranch near Lake Osoyoos. The local climate and soil were ideal for the growing of hard and soft fruits, grapes, and other produce. It was said of him that he never turned anyone away hungry, even if they had no money. And he was never happier than when his table, which seated twenty-four, was completely surrounded by visitors and friends. His trading post became the rendezvous for Indians, miners, priests, and trappers from various places throughout the region. The productiveness of his ranch was evidenced by the fact that in one season it produced 1,000 bushels of peaches. The place was a beehive of activity during harvest time. Both whites and Indians were employed with the picking, drying, and canning of fruit. A prospector who visited the ranch described peaches being dried on sheets, wagon covers, and all kinds of cloth laid on the ground. There was one evaporator in use. Smith sold his apples at not less than 6¢ a pound and always had a ready market. Four of these old, gnarled apple trees and one pear still remain where Smith planted them over 130 years ago. And the fruit is just as tasty as it was back in the pioneer days when such items were considered a rare delicacy.

Okanogan Smith had posted several claims prior to the establishment of the Moses (Columbia) Reservation in 1879. When the perimeters of the Moses

Of the trees planted by Smith around 1860, four apple and one pear are still living and producing fruit. The apple tree pictured is a Winesap.

Okanogan Mine

Notice is hereby given that I have this day located and claim fifteen hundred feet on this Lode. leads vein or deposit of Mineral bearing quartz. Commencing at this monument and running along the strike of the lode in a North Northwest direction five (500) hundred feet from this monument and running along the course or strike of the lode in a South East direction (1000) one thousand feet. I also claim (300) three hundred feet on each side of the middle of the vein or lode together with all leads, lodes, veins or deposits of Mineral bearing rock within the lines of my location. This mine is situated about one half mile from the Similkameen River on the North side of the same and about one and a half miles from the 49th parallel in what is known as the Copper Belt.
This mine shall be known as the Okanogan Mine on the 49th parallel lode.

H. F. Smith

Mt Chopaca and Similkameen Mining District. Sept. 15th 1878

Official Seal

~~Recorded October~~
H. F. Smith
Recorder

State of Washington } ss
County of Okanogan }
I hereby certify the above to be a true copy of the record of the Okanogan Mine as presented to me by the record book of the Mt Chopaca and Similkameen Mining District.

Chas. E Sears

(Left) Notice of claim and (below) application for survey of the Okanogan Mine. (Courtesy of Dale DeVon)

Application for Survey of Mining Claim.

Okanagan Mine.

Claimed by H. F. Smith and Marshall Blinn

In Similkameen Mining District,

In Stevens County,

Township No. not surveyed Range

~~No.~~ East of the

Willamette

Meridian, Washington Territory.

Dated June 9th 1884.

Filed June 9th 1884.

Estimate ($ 35.) Sent to

H. F. Smith: Olympia. Thurston

County, Wash. Ter. June 9th. 1884

Instructions sent to Chas. E. Sears.

U. S. Deputy Mineral Surveyor,

June 16th, 1884,

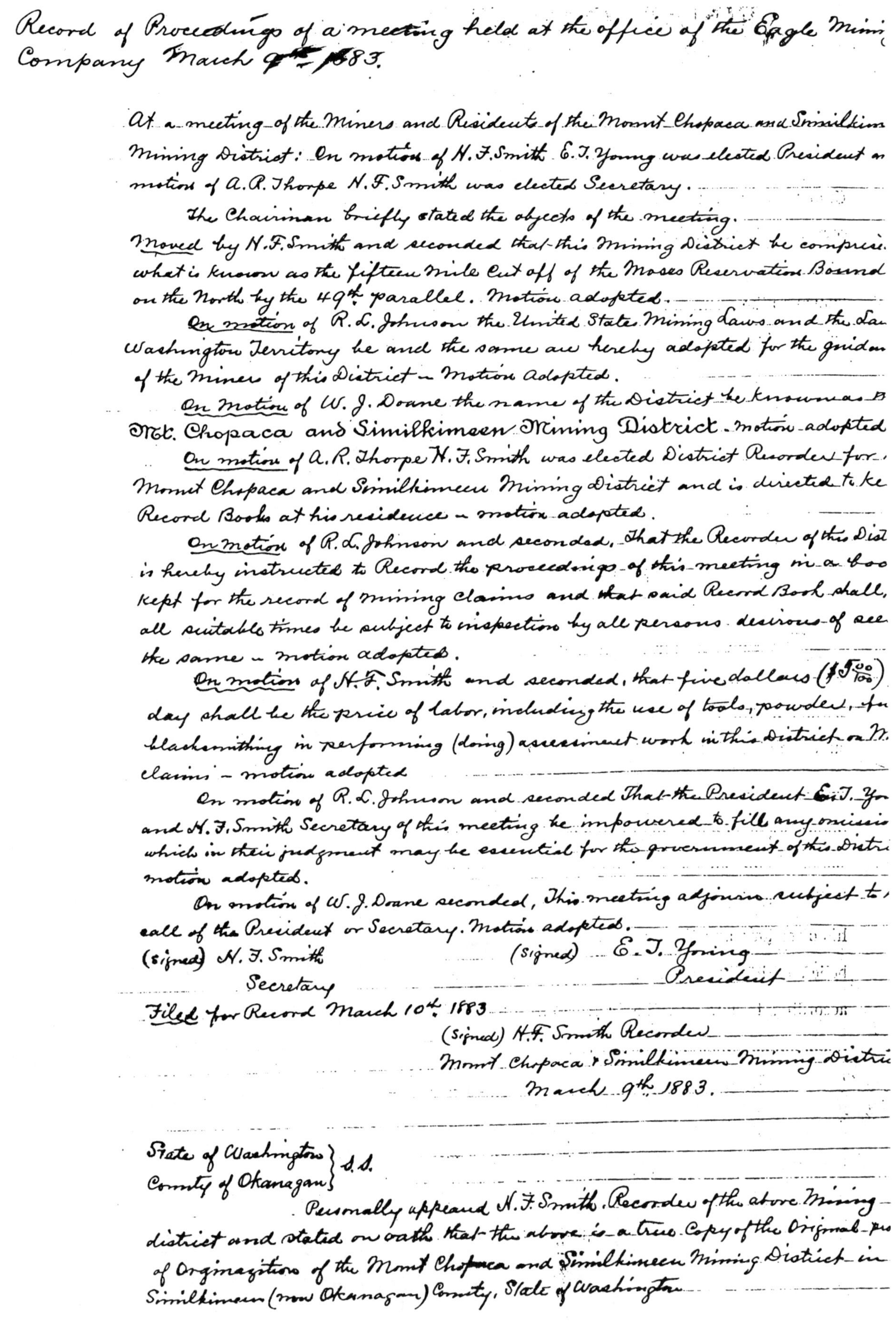

Record of Proceedings of a meeting held at the office of the Eagle Mining Company March 9th 1883.

At a meeting of the Miners and Residents of the Mount Chopaca and Similkimeen Mining District: On motion of H. F. Smith E. T. Young was elected President on motion of A. R. Thorpe N. F. Smith was elected Secretary.

The Chairman briefly stated the objects of the meeting.

Moved by H. F. Smith and seconded that this Mining District be comprised what is known as the fifteen mile Cut off of the Moses Reservation Bound on the North by the 49th parallel. Motion adopted.

On motion of R. L. Johnson the United States Mining Laws and the Laws of Washington Territory be and the same are hereby adopted for the guidance of the Miners of this District – Motion adopted.

On Motion of W. J. Doane the name of the District be known as Mt. Chopaca and Similkimeen Mining District. motion adopted

On motion of A. R. Thorpe H. F. Smith was elected District Recorder for Mount Chopaca and Similkimeen Mining District and is directed to keep Record Books at his residence – motion adopted.

On motion of R. L. Johnson and seconded, That the Recorder of this District is hereby instructed to Record the proceedings of this meeting in a book kept for the record of mining claims and that said Record Book shall at all suitable times be subject to inspection by all persons desirous of seeing the same – motion adopted.

On motion of H. F. Smith and seconded, that five dollars ($5.00/100) day shall be the price of labor, including the use of tools, powder, and blacksmithing in performing (doing) assessment work in this District on M. claims – motion adopted

On motion of R. L. Johnson and seconded That the President E. T. Young and H. F. Smith Secretary of this meeting be impowered to fill any omissions which in their judgment may be essential for the government of this District motion adopted.

On motion of W. J. Doane seconded, This meeting adjourns subject to call of the President or Secretary. Motion adopted.

(signed) H. F. Smith (signed) E. T. Young
Secretary President

Filed for Record March 10th 1883

(Signed) H. F. Smith Recorder
Mount Chopaca & Similkimeen Mining District
March 9th 1883.

State of Washington } S.S.
County of Okanagan }

Personally appeared H. F. Smith Recorder of the above Mining district and stated on oath that the above is a true Copy of the Original proceedings of Orginazation of the Mount Chopaca and Similkimeen Mining District in Similkimeen (now Okanagan) County, State of Washington

Handwritten record of the proceedings from the meeting held to organize the Mount Chopaka-Similkameen Mining District, March 9, 1883. H. F. Smith was elected secretary and district recorder. (Courtesy of Dale DeVon)

Reservation were established, it included his mining properties in the Similkameen Valley. Confronted with the threat of having to forfeit his claims, he offered to sell them to the government for the sum of $250,000. No payment was granted. To discuss the problem, Smith chaired a meeting of local miners and cattlemen who, like himself, had a lot to lose. The outcome was a request of the government to exclude a strip of land along the north end of the reservation. As a resolve to the matter, President Arthur in February, 1883, issued an executive order restoring to the public domain a fifteen-mile strip of land fronting to the north on the British Columbia boundary line and running east to west from the Okanogan River to the 44th degree of longitude west from Washington, D.C. On March 9, 1883, at the Eagle Mine, a meeting was held in which the restored section was organized into the Mount Chopaka-Similkameen Mining District. In 1886, the remaining portion of the Moses Reservation was thrown open to mineral entry.

Smith served in both the territorial and state legislatures. While a member of the territorial legislature, he introduced a memorial to Congress requesting that there be obtained from Russia certain fishing privileges along the coasts of Russian America. The memorial, addressed to President Andrew Johnson, was adopted by the legislature of Washington Territory on January 13, 1866. When the memorial from the territorial legislature came into the hands of the Secretary of State, William H. Seward, it prompted negotiations which led to the purchase of the territory from Russia. [11]

Powerhouse at the "falls" on the Similkameen River. (Photo courtesy of Harry Sherling)

While serving in the state legislature, Smith became acquainted with a news correspondent by the name of James Hagerty. Hagerty was likely captivated by Smith's stories of the limitless wealth which lay beneath the surface of the far-off Chopaka-Similkameen mining district. So it probably didn't take much to convince the reporter to leave when asked by the Tacoma Ledger to investigate mining activities in the Okanogan. In 1892, at the newly-settled town of Oro (Oroville), Hagerty started a newspaper, the Madre D'Oro, which he managed for a short time. Both men had spent time in the metropolitan centers and were well informed of the latest inventions. They very likely had seen the electric streetcars of the big city and observed how they worked. The question must have arisen - if this invisible energy called "electricity" could turn the wheels of trolley cars, then why not harness it to work the mines in the Okanogan? The January 8, 1892, Okanogan Outlook, reported Smith's interest in the development of electrical power:

> *Representative H. F. Smith is working on a scheme to irrigate the Okanogan Valley, furnish electric power to operate and light the mines in the vicinity of Palmer Mountain and Osoyoos, to run a large smelter, which is to be built on the Similkameen, and connect the principle mining camps of the country by electric cars. The scheme which involves several million dollars is progressing rapidly towards maturity, and the promoters of the enterprise expect to have everything in shape for business as soon as spring opens.*

After reporting on the Klondike gold rush in 1897, James Hagerty moved to Nighthawk where he continued with the development of some of Smith's mining claims. Smith had died in 1893. The extent to which Smith and Hagerty had shared the dream of implementing electricity is unknown. However, it is known that Hagerty promoted the building of a power plant on the Similkameen. In June, 1905, the small plant was being brought to completion. Lumber for the powerhouse was to be delivered the next month. When finished, it supplied electricity to

Oroville and the more important mines of the district. It was a precursor to the Enloe Power Dam built later at the same location. The Enloe Dam and powerhouse site, because of its historic significance, is listed on the National Register of Historic Sites.

How did Hiram come to be called "Okanogan?" Other politicians probably nicknamed him "Okanogan" in recognition of the fact that he represented that portion of the state. Another opinion is that he was first called "Okanogan" Smith by the Indians because the place where he settled was called "Okinikane," meaning the "place of rendezvous." Indians came from distant places to meet each year at the head of the Okanogan River for a Potlatch. At this location they stored up a reserve of fish and game for the coming winter months. The original term, Soyoos, which meant "narrows" was corrupted to Osoyoos. This was known to be one of the best fishing spots in the Northwest. [(12)]

Smith was elected to serve in the State House of Representatives (the second legislature) for Okanogan County, and then, in 1892, as the incumbent (Democrat), defeated the former lieutenant governor, Charles E. Laughton (Republican), to serve as a member of the third House of Representatives. In 1893, after the adjournment of the legislature in the spring, Smith went to Seattle to rest and recuperate from a lengthy illness. On September 9, while staying at the Diller Hotel, Hiram "Okanogan" Smith passed away. He was 64 at the time. Being a well-known pioneer and statesman, his death was mourned by many. The humble printer-turned prospector had earned his place among the notables of early Washington State history.

The September 10, 1893, issue of the Seattle Post Intelligencer announced the death of Hiram F. Smith:

> *. . . one of the oldest and best known pioneers of the Pacific Northwest, died of dysentery at 10:15 o'clock last night at the Diller Hotel where he was temporarily stopping. . . . by the death of "Okanogan" Smith the state of Washington loses a whole-souled, generous and energetic citizen. With the development of the Okanogan County, Mr. Smith had more to do than any other man, probably. He was the one man who had the courage to go into the wilderness and make it possible for a white man to live in the Okanogan country. For twenty-eight years or more his farm and store or trading post at Osoyoos Lake have been known as the great center of the whole Big Bend country. . . . The immediate cause of Mr. Smith's death was a cold caught while out on lake Washington some time ago. . . .* [(13)]

Hiram Smith was laid to rest at the Lake View Cemetery in Seattle on September 12, 1893, near the grave sites of Henry L. Yesler and Otto Ranke. Three senators and three representatives from the state legislature served as pall bearers. Though Smith was buried at Lake View, it remains a mystery as to the exact location of his grave.

THE BLEWETT MINING DISTRICT

Peshastin Creek flows along present-day Highway 97 which travels down the north slope from Swauk Pass and through the historic site of the Blewett mining camp. Except for an information sign, little remains to indicate where the town of Blewett was once located.

For decades, writers have embraced the belief that placer gold was discovered along Peshastin Creek around 1860 by prospectors returning south from the Similkameen River and Cariboo district of British Columbia. The date, 1860, coincides with the year that gold was discovered on Ingalls Creek, a tributary of Peshastin Creek. Ingalls Creek empties into Peshastin Creek about five miles down the valley from the Blewett townsite.

As announced by the September 14, 1860, Pioneer and Democrat, gold was discovered on Ingalls Creek by a group of about fifteen miners on their way to Rock Creek in the Okanogan highlands. The discovery of gold on Rock Creek the previous fall had prompted a migration of miners to that area. During the spring and summer of 1860, Noland & Company was reported to have taken out $7,000 from Rock Creek, $700 in one day. A nugget was found weighing $32. A letter received from Colville in August, 1860, mentioned that some of the miners who had gone from Colville to Rock Creek had returned for their friends and that many were doing well. There were some, however, who were not making expenses. One thousand miners were expected to remain in the vicinity of Rock Creek during the winter of 1860-61.

When the fifteen miners bound for Rock Creek discovered gold on Ingalls Creek, they decided to go no further. They dropped off their loads, then sent the pack train back to Seattle for more supplies. Gold samples were taken along to be delivered to friends of the miners. When the discovery of gold on

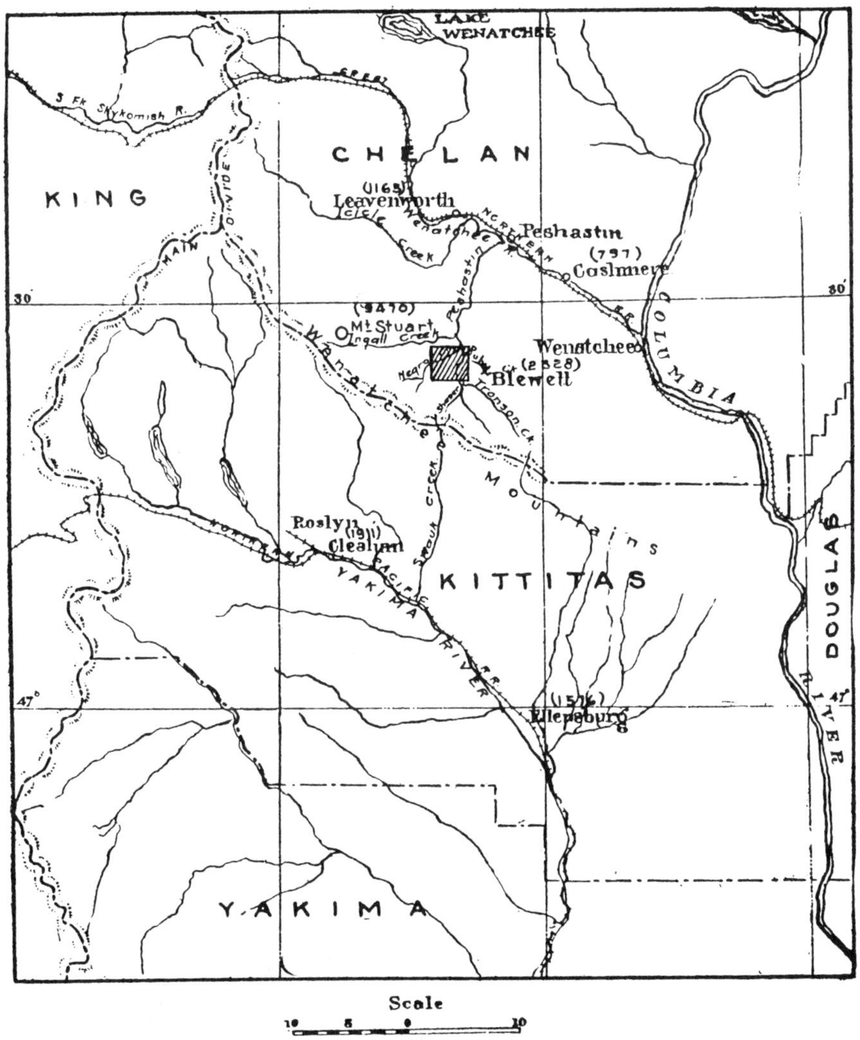

MAP OF EASTERN PORTION OF CASCADE MOUNTAINS BETWEEN THE WENATCHEE AND YAKIMA RIVERS

Area Representing Blewett Mining District Areal Geology Mapped

Map of eastern portion of Cascade Mountains between the Wenatchee and Yakima Rivers from the Washington Geological Survey, Bulletin No. 6., Plate II, 1911. (14)

Ingalls Creek was made known in Seattle, there was a great celebration "with firing of guns, colors flying, and a general rejoicing." The news of the discovery had a profound economic influence on the young town. Building lots that previously sold for $100 took a shot upward, ". . . to an almost unwarrantable price." Yesler and Rand's steam sawmill was temporarily shut down because the head sawyer, engineer, and all hands had taken off to travel to the new El Dorado on Ingalls Creek. Thomas Russell, a well-known former sheriff of King County was on his way to Rock Creek to deliver the express when he stopped at Ingalls Creek to prospect. He sent the express on with his brother, Loe, and wrote to his friends in Seattle that with his new "lay out" he would carry no more express. Russell found from 3¢ to 10¢ to the pan in the banks of the stream. Some of the miners reportedly were getting as high as $1 per pan. At that time, there were about ninety men on the creek. Russell began building sluice boxes and hoped to begin "raking in" the gold the next day. [(15)]

It was probably about the time of the stampede to Ingalls Creek that an African American worked the bars of another creek which emptied into Peshastin Creek, a short distance upstream from Ingalls Creek. He was reported to have taken out $1,100 in one season.

The first quartz ledge discovered in the Blewett District was reportedly by a United States soldier in 1854. The location was never recorded. In 1874, John Shafer located a quartz vein near the head of Culver Gulch, which became known as the Culver claim. In December, 1891, L.E. Grimes of Ellensburg gave a brief history of the Culver and other mines in the Blewett District:

> *The Culver mine, formerly known as the Shafer, was located by Ira Canady about 14*

The town of Blewett at the turn of the century. The 20-stamp mill is located at the right and a hotel can be seen in the distance. Cordwood has been stacked to use as fuel for the steam engine. (Photo from Washington Division of Geology and Earth Resources files)

Blewett miners gather from the surrounding hills to pose for a picture. Notice the miners' candlesticks which were used in the mines to hold candles as a source of light. The hand-pushed ore car was used in many of the early mines. A trip-lever released the door at the front allowing the ore to roll out as the car was being tilted forward. (Photo courtesy of the North Central Washington Museum, Wenatchee)

years ago. Mr. Canady sold it to Thomas Johnson, who took about $100,000 from a pocket and then abandoned it. William Donahue relocated the claim, sunk a shaft, and ran a crosscut and, after considerable litigation, sold the claim to Warner and Bash for $15,000, $5,000 cash and $10,000 when the mill was put in operation.

The mine is rich in free-milling ore that averages $50 to the ton. A twenty-stamp mill has been shut down until spring. The improvements are many, including a three-mile endless cable which covers a mile and a half from the mine to the mill. When Mr. Donahue sold the Culver, he retained an interest in the Manastee and the Peshastin. From the latter claim he ground out $80,000 in an old arasta (sic) last spring and packed the product out on Cayuses.

Another gentleman who is operating extensively in this camp is John A. Shoudy, mayor of Ellensburgh, who with John Earnest has struck free-milling rock on the Black Jack, Pole Pick and extensions, which assays about $40 per ton, and a stamp mill will be put up in the spring. [16]

Author's note: (the "h" in Ellensburgh was dropped in the 1890's)

The rush of miners into the Blewett District gave rise to the Blewett mining camp. As a travel route to the outside, a wagon road was built in 1879 connecting Blewett with Cle Elum to the south. In 1898, a stage road was built connecting Blewett with the town of Peshastin to the north. During its peak years the town supported a population of several hundred people, but after the turn of the century dwindled to only about forty. It's been estimated that from the time of the first gold discovery until 1910, the district produced about $1,700,000. [17] Most of this came from the mines in Culver Gulch, among which were the Pole Pick, Hummingbird, Bobtail, Fraction, and Little Culver.

When ore production in the area exceeded the capacity of the arrastras, stamp mills were built to process the gold. In 1892, the Blewett Company built a twenty-stamp, steam-powered mill at the mouth of Culver Gulch, allowing space for twenty more stamps. Wood was burned under two boilers (used alternately) to provide steam for a 50 h.p. Corliss engine. Other features were described as follows:

> *"The batteries consist of four sets of five stamps each, with Fraser and Chalmers automatic feeders, and double discharge, only single being used. 950 pound stamps, with chrome steel shoes and dies, drop 6½ inches at the rate of 90 per minute ... A Hallidie aerial tramway with buckets holding 250 pounds each, carries the ore from the mine 4,000 feet distant and dumps it into two receiving bins of 400 tons capacity. The usual system prevails in regard to the different floors of the mill, the order here being crusher, feeding bin, battery, and concentrating floors." (Professor Milner Roberts quoted.) During the year of 1896 this mill is said to have reduced 2,469 tons of ore from the*

Miner's candleholder and carbide lamps. Calcium carbide used in carbide lamps was first discovered in 1862, then accidentally rediscovered by T.L. Wilson of North Carolina in 1892. The first commercial production was in 1896. Carbide lamps probably came into use shortly thereafter. The lamp was worn on a miner's hat as shown. A water reservoir at the top supplied a regulated amount of water to activate carbide (below) into combustible acetylene fumes. The large lamp on the right is a "shift boss" lamp. The candleholder (also called candlestick) was in use prior to the invention of the carbide lamp. It has a sharp end which was driven into a timber or small crack in the sidewall of a tunnel. Miners' candleholders are considered valuable collector's items.

Culver claim, which assayed $12.60 a ton, and in addition to this 473 tons of customs ore; the total product of ore in bullion for the company in the year 1896 being $60,000.[(18)]

All that remains of the mill today is a portion of its foundation and a few standing timbers which can be seen on the hillside a short distance from Highway 97 north of Swauk Pass in Washington State. The original foundation stones were laid in place without the use of mortar. Unfortunately, vandalism has taken its toll on this valuable historic landmark.

The February 16, 1889, Kittitas Localizer of Ellensburg described the mills at work along Peshastin Creek:

Crossing over the head of the Swauk comes the Peshastin country where are located a number of gold claims and 2 mills. The ore from this district is free-milling and some of it runs $50.00 to the ton. Fifty pounds of amalgam retorted at one of the mills gave a result of 29% of metallic gold, making a clean up for one week's work of $3,625.

A few timbers and foundation stones are all that remain of the twenty-stamp mill at Blewett.

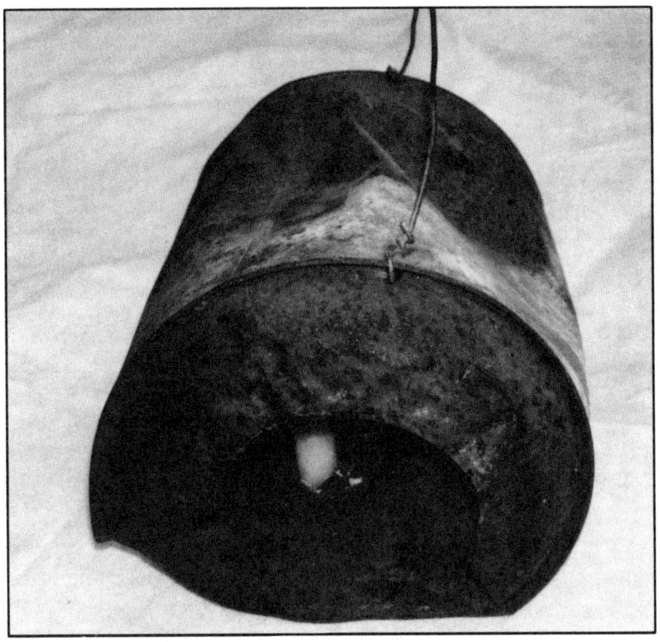

The "bug light" was contemporary with the candleholder. The light was made from a can with a large open end, such as a baking powder or lard can. For ventilation, small slits were punched around the perimeter of the end opposite the opening. As a source of light, a candle was held in a hole in the bottom of the can (when used as a light, the side became the bottom). The hole was made with two intersecting cuts. The sharp triangle corners of the hole kept the candle in place. A wire bail was attached to the top. When the lamp was being used, the light shinning through the slits in the back created the illusion of "bugs" on the snow. These were carried by early miners while walking to and from the mines and sometimes in the mines. And, of course, the light was used to find one's way to the outhouse. The bug light was homemade and never produced commercially, therefore, as a collector's item is rarely found. In later years, the term "bug light" was slang for a miner's electric cap lamp run from a battery.

MILLING MACHINES

Mills were designed to reduce ore to concentrates from which precious metals were then extracted. Blocks of stone, jaws, balls, rollers, wheels, and pestles were some of the instruments used to crush ore. The first Greek and Roman ore crushing devices were hand-mortars and millstones similar to those used in grinding flour. The procedure was improved upon with the invention of the stamp mill sometime during the late 15th or early 16th centuries. A stamp mill described in Europe in the mid-1500's had sophisticated features, some of which resembled those of American versions built during the 19th and 20th centuries:

> *. . . these cross-beams have square openings, in which the iron-shod stamps are inserted. The stamps are not far distant from each other, and fit closely in the cross beams. Each stamp has a tappet at the back, which requires to be daubed with grease on the lower side that it can be raised more easily. For each stamp, there are on a cam-shaft, two cams, rounded on the outer end, which alternately raise the stamp, in order that, by its dropping into the mortar, it may with its iron head pound and crush the rock which has been thrown under it. To the cam-shaft is fixed a water-wheel whose buckets are turned by water-power.*[19]

Mill buildings in America were usually built on hillsides to take advantage of gravity in implementing the various stages of the gold recovery process. Concentrated ore particles were channeled over amalgam tables or other separation devices such as the Wilfley table.

MILL CONSTRUCTION

SIDE ELEVATION

DESIGN FOR 10-STAMP AMALGAMATING-CONCENTRATING MILL, WOOD CONSTRUCTION, STEAM DRIVEN

Material required:—68,000 feet B. M. Lumber. 53 squares corrugated iron roofing. 70 squares building paper for sides and ends. Rods, bolts and washers, 1500 pounds.

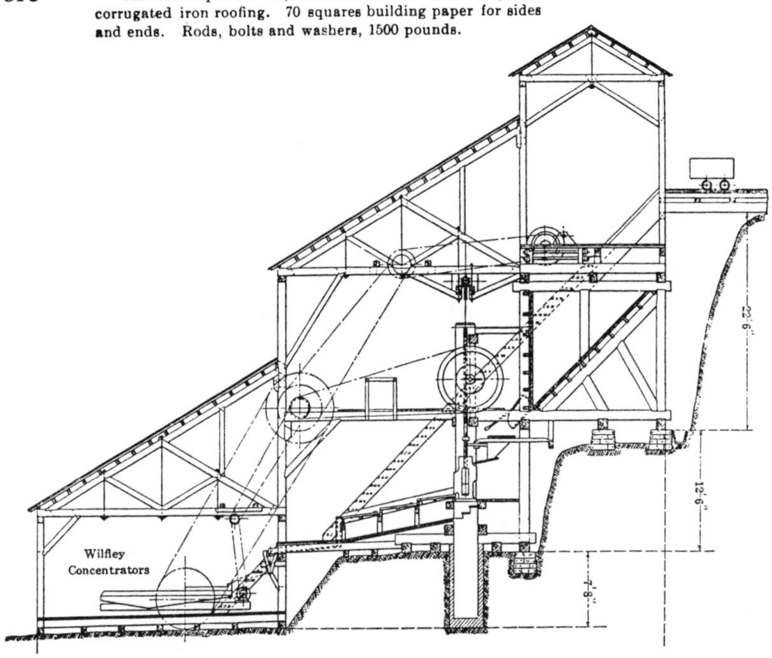

Ten-stamp mill advertised in The Mine and Smelter Supply Catalogue, 1912. The stamp mill was an apparatus in which rock was crushed by descending pestles (stamps), and operated by water or steam power. The stamps were generally grouped in units of five per mortar. They varied up to 2,000 pounds in weight, dropping six to eight inches and over 100 or more times per minute.

The Lane Mill was a slow-speed roller mill of the Chilean type. Six roller wheels revolved slowly in a pan ten feet or more in diameter making about eight revolutions per minute. The mill shown was used to crush ore at the Eureka (New Light) mine in the Slate Creek Mining District at the turn of the century. Crushed ore was washed from the trough for processing. The mill was removed from its original location and reassembled at the Shafer Museum in Winthrop, Washington, where it is on exhibit.

An early mine and mill site was recently discovered at an undisclosed location on the west slope of the Cascades. It is several miles from a major road. A number of milling machines were found at the site. This arrastra-like concentrating mill was patented in 1898. It was called "The Montgomery Ore Granulator and Pulverizer." For the grinding of ore, it used three large steel balls, each about twenty inches across (laying on the ground). Each ball is estimated to weigh around 1,000 pounds. The balls and other heavy components had to be carried many miles up steep terrain, through heavy forest, and across a river and several streams. It would be interesting to know exactly how the miners transported these burdensome objects to the mountainside where they now rest. Mules or horses were undoubtedly used. Historians are interested in preserving these valuable artifacts.

A Pelton water wheel discovered at the abandoned mine site. The wheel probably drove a generator and other mill machinery. Pelton wheels, used to harness water power, were a common fixture at early mines. A few of these heavy iron wheels still remain where they were left nearly 100 years ago.

Next to the Montgomery Ore Granulator is a jaw crusher entangled in nearly a century of vegetation growth. The flywheels (as can be seen) carried the momentum of the jaw's crushing force. Shattered ore from this crusher was reduced further by the grinding of the ore granulator. Amalgamation plates or an ore dressing table (possibly both) were probably installed to separate valuable minerals from worthless gangue. Many early mines used amalgamation (copper) plates coated with mercury (quicksilver) to gather gold particles from crushed free-milling ore. The resultant gold-mercury alloy was called "amalgam."

THE ARRASTRA

The term arrastra or "arrastre" (Sp. Arrastrar) carries the meaning "to trail along the ground." The arrastra, also called the drag mill, was an early Spanish-American device used for pulverizing ores containing free gold. Powered by water wheels, horses, or mules, the mill had one or more drag stones which were pulled in a circular direction over chunks of ore. The ore was crushed to a fine powder, thus freeing the gold particles which were then gathered with mercury. The resultant alloy mixture was called "amalgam." To separate the gold from the mercury, miners used the retort method. This involved heating the mixture to vaporize the mercury, leaving "sponge gold" at the bottom of a specially constructed retort. As the mercury vapors traveled upward through a tube, they were cooled and condensed into original consistency. Retorting the mercury in this manner allowed the miners to reuse it. In addition to using mercury in the recovery process of the arrastra, miners used it to gather gold in an assortment of other devices including copper-bottomed pans, sluice (riffle) boxes, barrels, and on copper plates. Miners and others became aware of the hazards associated with the improper use of mercury, especially as regards vapors. Bloxam's Chemistry, 1890, mentions the suffering experienced by operatives (those making mirrors and barometers) from the poisonous action of mercury.

While on the Forty Mile River in Alaska, we encountered amalgam in an unusual form. A dredger came down to our camp to run his concentrates over the spiral wheel. We watched with interest as small silvery beads rolled up the terraces toward the hole at the center. It was soon discovered that the beads were gold fines suspended in mercury. A discussion ensued as to the origin of the mercury, whether natural or the product of early mining. One dredger argued for the "natural" theory because he had found

Old arrastra on Swauk Creek, Kittitas County, as it appeared in 1924. The mill was powered by an overshot water wheel. Stones were pulled in a circular motion over free-milling chunks of ore, crushing them to powder. (From Moen, 1982, p.7) [20]

mercury in an undisturbed (unmined) layer of gravel. In either case, the mercury had bonded with the gold - an unusual occurrence.

After the turn of the century, in the Blewett District, two abandoned arrastras were still intact near the mouth of Culver Springs Gulch and remnants of others were located on Peshastin Creek and in Culver Gulch. The one best preserved was described as follows:

> *The pit is twelve feet in diameter, three feet deep, and built of granite from outcrops not far away. The power is obtained by a large overshot water wheel, twenty-six feet in diameter, and run by a stream of water brought from the gulch by a flume and carried directly over the upper portion of the wheel. This operates a horizontal wheel to which granite boulders weighing over one-half a ton are attached by chains. . . . The arrastra is said to have crushed from one to two tons of ore per day.* [21]

Arrastra stone base located next to Highway 97 north of Swauk Pass in Washington State. Drag stones used to crush ore wore a round, smooth trough into the stone base.

POWER DRILLING

Power drills were usually driven by compressed air. Of those driven by air, there were two types - the piston drill and the hammer drill. The first to be developed was the piston drill. With this type of drill, the steel became an extension of the piston and moved back and forth with it. The hammer drill differed in that the piston acted as a hammer striking the head of the steel.

Stoper and drifter drills were of the hammer type. The stoper was supported on a pneumatic jackleg which applied forward pressure to the drill steel. This make of drill was used in tight areas and for drilling upwards. The drifter was much heavier and was used more for driving drift tunnels. It was supported on a vertical or horizontal column wedged between the walls of the drift. The wet drill, sometimes called the "waterliner", provided the added feature of water which was propelled through the shank of the drill steel. Water helped keep the steel cool, clean the hole, and hold down the dust. The dry version did not use water and was considered hazardous to miners because of the dust it created, resulting in silicosis. It was nicknamed the "widow-maker."

In driving a drift, a series of dynamite holes were drilled at various levels into the face of the rock. Graduated lengths of bits were used to reach the desired depth. Whenever the bit got stuck, the "Chuck Tender" (miner's helper) struck the bit on the side to jar it loose. It was the job of the "Nipper" to deliver freshly sharpened bits to the mine and dull ones to the blacksmith.

As remembered by my friend, Harry Sherling, who mined as a young man, drill holes were arranged in a specific pattern. Very near the bottom of the face, "lifter" holes were drilled into the rock. Above the lifter holes, "cut" holes were drilled in the downward direction. Just above the "cut" holes, about midway up the face, "breast" holes were drilled. At the upper part of the face, near the roof, "back" holes were drilled. To reach the back holes, the horizontal bar supporting the drill had to be moved upward.

The cut holes were the first to blow, then the breast holes, followed by the back holes, and finally the lifter holes. Each hole was timed by fuses of different lengths. Prior to the blasting of the holes, a sheet of steel was laid on the floor of the tunnel to make shoveling easier for the muckers. Harry says it was a "rushed deal." As soon as the muckers moved out, three more brought in another car to be loaded. Even before the muck pile had been cleared, miners were already moving into position to begin drilling the back holes near the top.

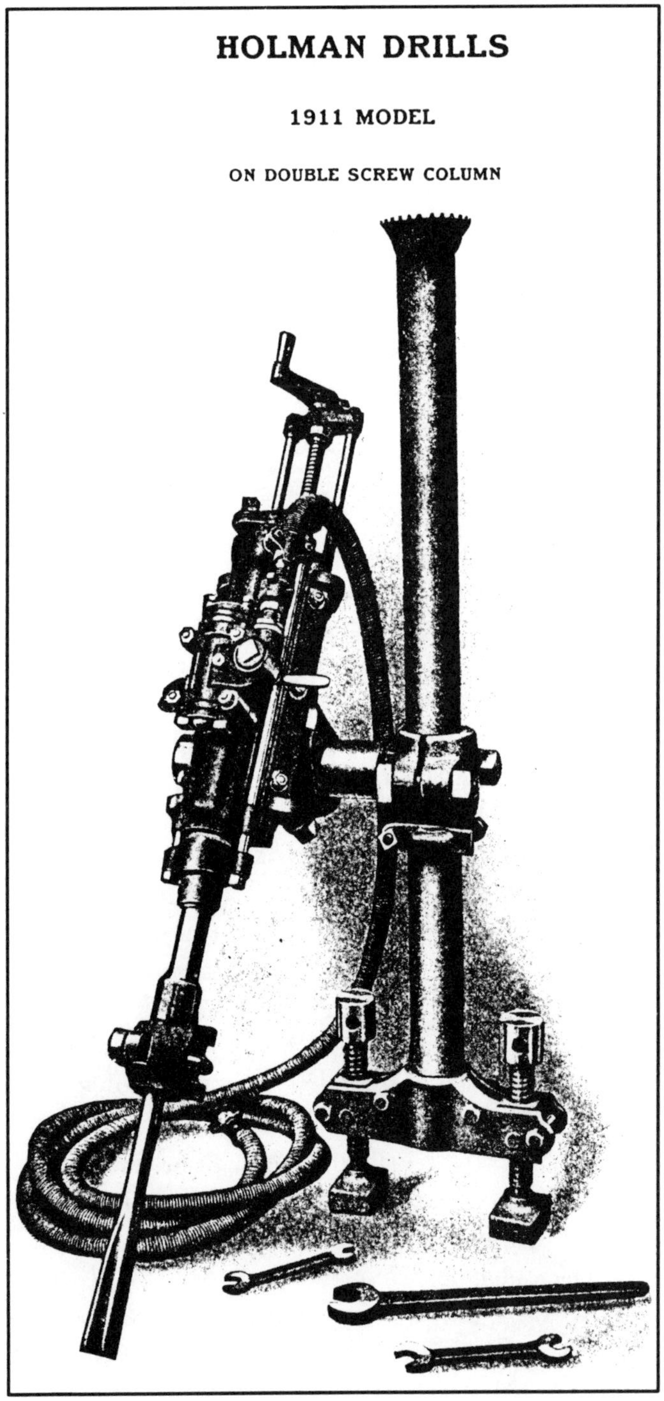

Holman drill advertised in The Mine and Smelter Supply catalogue, 1912.

Dynamite was carefully tamped into each of the holes with a wooden rod. Usually dynamite of 40% nitroglycerin was used. If the rock was especially hard, 60% was sometimes preferred. The shattering power of dynamite becomes stronger with the increase of nitroglycerin. When dynamite was in place in the hole and the fuse with detonators in position to set off the primer stick of dynamite, miners used their carbide lamps to start the fuse burning. The end of the fuse was split to expose the powder train. To light the fuse was to "spit" the fuse. When the fuse was ignited, a flame shot out the end creating a hissing sound. This was the spit. Sometimes miners used a "spitter" which was a length of safety fuse with gashes cut into it every few inches. As the fuse burned at each gash, it shot out a hot flame which was used to light the fuses going to the various holes. If the round had nine shots, then at least nine slashes were cut. The spitter could also double as a timer. If the miner started with a five-foot fuse to the first hole and wanted two feet of it unburned when he was finished lighting the rest of the fuses, he would select a three-foot section of spitter fuse. When the three-foot spitter fuse was completely burned and, hopefully, all the fuses had been lit, he knew he was running out of time and had better get out of there. The blasts in a round were always counted. If for some reason a charge was delayed and did not go off, a safe margin of time was allowed before re-entering the tunnel. Miners also had to be careful in handling muck which might contain dynamite blasted free from the face.

A procedure sometimes used in the preparation of a hole for blasting was "springing the hole." Prior to setting off the main charge, a smaller charge of one or two sticks of dynamite was used to enlarge the bottom of the hole so that more dynamite could be tamped into it, creating a larger blast and requiring fewer holes to be drilled. For the main blast, the sticks of dynamite were slit three times lengthwise allowing them to squash together when compressed into the hole. By doing this, the dynamite took up all of the space in the hole resulting in a more effective blast.

HAND DRILLING

Prior to the invention of the pneumatic hammer drill, miners drilled blasting holes into rock by use of the chisel-bitted drill steel and hammer. Even after the pneumatic drill came into use, hand drilling was used in many of the mines where power equipment was impractical. A three to four-pound "single-jack" hammer was used to drive the steel into the rock, usually two to three feet deep. The miner held the drill steel with one hand while striking it with the other. The steel was rotated between each blow to insure a round hole and to keep the drill from getting stuck. In this way, a hole could be driven in a remarkably short time. In drilling "up" holes into the roof of a drift, with practice a strong miner could swing the hammer in a full circle, striking (bumping) the drill steel on the way up. Jack Stewart, a mining engineer who lived in the Methow, had the reputation of being an expert hand driller and was especially skilled in drilling "up" holes. Jack had taught young miners the methods of hand drilling. In 1932, Vernon LaMotte was mapping the Flag mine near Mazama. He watched as Jack drilled a hole in the roof of a drift to hold a plug from which to suspend a plumb. Jack held the steel and turned it with one hand while he swung the maul in his other hand. LaMotte knew this was a technique used for overhead stoping (removing ore from above a drift), but had never seen it done. He and the others present were amazed to see such a feat, even by a large man over six feet tall.

In "double-jacking," the hole was usually driven six to eight feet. One miner held the steel while his partner used both hands to swing a heavier hammer, usually eight to sixteen pounds. In driving deeper holes, the lengths of steel used were progressively smaller in diameter to fit the taper of the hole, caused

An early makeshift blacksmith shop. The worker, probably at a small mine, is using a forge and blacksmith's bellows to sharpen a length of chisel-bitted drill steel. The forge may have been a "duck nest," a small cast-iron pot with an opening into which air was blown to intensify the heat. Rocks were used to contain burning charcoal. (Johanson collection)

by the wearing down of the chisel head. Drill steel had to be replaced with newly sharpened steel after being driven a foot or two, depending on the hardness of the rock.

When rock chips and dust gathered at the bottom of the hole creating a barrier for the drill steel, an especially designed spoon or scoop was used to clean it. The spoon was forged from a steel rod. The first few inches of the rod was shaped into a trough (elongated spoon with an end cap). It was shoved to the bottom of the hole, turned to scoop up the debris, then withdrawn. When drilling a "down" hole, water was periodically poured into the hole to suspend the cuttings in mud. To purge the mud, miners used either the spoon or a handmade suction pump. The pump consisted of a pipe inside of which was an extractor rod with a leather gasket. The gasket was attached near the end of the rod and held in place with washers. On a miniature scale, it worked similarly to the early domestic water pumps used in rural areas. The spoon was usually forged at the mine's blacksmith shop. The pump was also handmade.

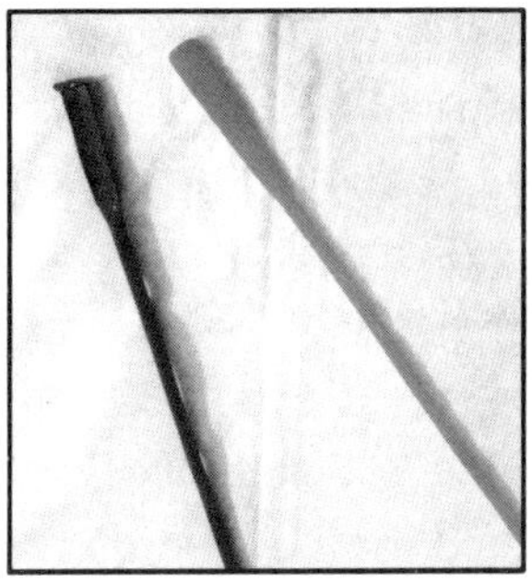

During the turn of the century, spectators gathered from miles around to see a rock drilling contest. Teams of two men competed to see which could drive the steel the farthest into rock in a fifteen-minute period of time. While one partner turned the steel, the other drove it with a double-jack maul. Every two minutes the partners traded places. Some partners were fast enough to switch without missing a stroke. Since water was used in the hole, a gunny sack was usually held around the steel to keep mud from flying into the face of the holder or bystanders.

John Berquist and John Gaffert were known throughout the Okanogan highlands as being very fast at rock drilling. To participate in a Fourth of July contest held at Chesaw, two miners came down from Rossland, B.C. Rossland was an active mining camp at the turn of the century. That day, the two visitors beat out all the competition and walked away with the grand prize of $75.00, considered to be a lot of money at that time.

Rock drilling contest in the town of Chesaw during the Fourth of July celebration, 1902. The Louie Larson log cabin in the foreground was the first building in Chesaw. (Photo courtesy of the Molson Museum)

THE POLAND-CHINA MINE

The Poland-China was one of several mines in the Okanogan that produced over $100,000 in gold. The original claims were staked in the spring of 1896 after rich ore was discovered near the surface. The mine was located on the North Fork of Mary Ann Creek in the Myers Creek District, about midway between the pioneer towns of Molson and Chesaw. The Poland-China assay office and other mine artifacts are on exhibit at the Molson Museum.

Different theories exist as to how the Poland-China received its colorful name. One is that two persons, one of Polish descent and the other of Chinese, held original interests in the property. A longtime resident of the area once saw a map which showed Mary Ann Creek as "China" Creek, possibly indicating the presence of Chinese. Also, Chinese influence in the area is indicated by the name Chesaw, an early mining town located on Myers Creek east of the Poland-China. The town derived its name from Chee Saw, a Chinese man who lived on Myers Creek with his Indian wife. They made their living from agricultural pursuits. The little cabin where they lived was on the trail through the north country used by Indians and a few white settlers. The trail went from the south end of Lake Osoyoos to Chee Saw's cabin, up to Rock Creek and Midway on the Kettle River in British Columbia, then downriver to the Colville vicinity. When the North Half of the Colville Reservation was opened to mineral entry in February of 1896, prospectors flocked to the Chee Saw crossing called "Chee Saw's ford" on Myers Creek. The town soon became a lively mining center with 250 inhabitants living in tents and log cabins. Today, Chesaw remains a small village with a few businesses and residences along its main street. In recent years there has been a renewal of interest in the potentially rich gold deposits discovered in the hills east of the town. As during the turn of the century, there is still an annual celebration on the Fourth of July. There are no longer rock drilling contests but rather a rodeo which draws people from various parts of the country.

John Grigsby, Barney Quinn, and Wiley Grigsby with tools of their trade at the portal of the Poland-China mine. (Photo courtesy of Richard Dart)

Inside the Gold Axe mine near Chesaw. A drifter drill is mounted on a vertical column. John Berquist is second from the left. The children have dressed for the occasion. (Photo courtesy of the Molson Museum)

THE PALMER MOUNTAIN TUNNEL

The Palmer Mountain Gold Mining and Tunnel Company was organized in 1895. The mine portal and 100-stamp mill were located on the southern slope of Palmer Mountain above the town of Loomis, Washington. The mill building, the largest ever built in Washington State, was 265 feet long and 104 feet wide. The mill was built with hopes that the mine would soon become a major producer of gold and silver. Early on, interest was generated when the manager of the mine, John Boyd, came down from the mountain with a chunk of ore containing free gold. The September 10, 1897, issue of the Palmer Mountain Prospector reported, "It is a good lead, assaying $185.20 to the ton in gold and $2.50 in silver. It was assayed by O. S. Stocker." [22]

Palmer Mountain Gold Mining and Tunnel Company mill building in the early 1900's.

As it turned out, though, the rich ore sample was not representative of ores within the mine. In the fruitless endeavor to unearth paying veins, a double track tunnel was driven more than a mile into the mountain, mostly through solid rock. Diamond core drills were used to broaden the search, but this, too, produced less than favorable results.

Palmer Mountain became the focus of attention in 1890 when a huge nugget was said to have been found somewhere in its terrain. Circumstances of its finding are vague but it was rumored that the nugget brought its finders $1,000.

During the peak of operation, about 200 men were employed at the Palmer Mountain mine. When anticipated gold reserves were not found, investors withdrew their support and the mine was subsequently closed. The huge mill, fully equipped for operation, never turned a wheel.

As related to the Palmer Mountain Gold Mining and Tunnel Company, an interesting story surrounds the person of James Hagerty. During an interview by the Spokesman-Review in Spokane in 1901, he made some disparaging comments regarding the lack of gold in the Palmer Mountain Tunnel. He deposited $50.00 with the editor of the Spokesman-Review and promised that if the newspaper sent an expert who could find ten ledges, each four feet in width, with values of $10.00 per ton, he would forfeit the money. If he were to be proven wrong, in addition to the $50.00, he would pay for the expert's expenses.

When the miners back at Loomis received word of Hagerty's untimely statements they became quite angry, knowing this could jeopardize continued investments and their jobs. Before his arrival back in town, a horseman was sent southward toward Conconully to make sure he was on the incoming stage. The horseman then hurried back to town to notify his conspiring buddies. When Hagerty arrived, he was met at the stage, then accompanied to a meeting hall where about seventy-five men had gathered. An inquiry was held in which he defended himself and added that if he were proven wrong, he would be willing to retract any statement. As they were leaving the building, he was thrown to the ground and warm tar was poured on his head, followed by a sprinkling of chicken feathers. After his assailants released him, Hagerty went to the Central Hotel to clean himself up and continue his journey home. One of the Riste brothers, who were then owners of the Golden Zone Mine at Chopaka, gave him a ride to his home at Nighthawk. Hagerty had built a hotel in Nighthawk around the turn of the century. It still stands today. Angered by the whole affair, Hagerty threatened to sue the ringleaders but apparently never did.

The tarring and feathering of Hagerty is a story well known among local historians. The consensus is that he was tarred and feathered for telling the truth.

James Hagerty

Room of 100-stamps inside the huge mill building. (Photo courtesy of Harry Sherling)

INVESTMENTS IN MINING

Huge amounts of money changed hands during the 19th century gold rush days. At a time when flour sold for around $3.00 a hundred, a pair of work shoes for $3.00, and bacon at 14¢ a pound (city prices), thousands of dollars were spent on speculative mining ventures. Well-to-do financiers provided capital for exploration and development. In one transaction in 1891, Messrs. Colby and Hoyt, Everett investors and directors of the Northern Pacific Railroad, were reported to have purchased two-thirds interest in the Monte Cristo mines - Pride of the Mountains, Pride of the Woods, Eighty Nine, and Galore Mines for the sum of "one dollar and other valuable considerations," as stated in the deed. It was learned from a reliable source that the "considerations" were actually $375,000. Payments included $50,000 up front followed by a cash payment of $150,000, and finally $175,000 to be paid from the earnings of the mines. The mines were only partially developed at the time. This was a fabulous amount of money to invest in a mining district which had been discovered only two years earlier. (23)

Courtesy of Center for Pacific Northwest Studies, W.W.U.

Small investors were encouraged to purchase shares of mining stock. Usually $1.00 or less (sometimes a penny) would secure the ownership of a share, thus giving the purchaser the hope of multiple returns and the added feeling of prestige in becoming a company shareholder. Some miners traded a portion of their labor for shares of stock. It was not uncommon to have several stock certificates among household documents. Most of these today are considered collector's items - a throwback to a time when a dollar represented a third of a day's labor. A good day's wage for a hardrock miner at the turn of the century was about $3.00.

W. WALLACE RADCLIFFE, President and General Manager.
GEO. GOE, Mining Expert.

The Puget Mining and Brokerage Co.

CAPITAL . . . $25,000

Office, 208 Pioneer Building, Seattle, Washington.

Dealers in Mines and Mining Stocks of Sterling Worth.

Correspondence solicited. MINES examined and reported upon. Information furnished by this company will ever be reliable in every detail.

We are in perfect touch with the prospectors in the best mining districts of Washington and British Columbia.

We are well posted on the most promising properties in course of development. This information we impart to our correspondents to their great advantage.

We are in constant correspondence with the leading capitalists and brokers in Eastern and foreign money centers. We will Buy, Bond, Develop and sell Mines and Mining Prospects of sterling worth.

We bring the prospector and the capitalist together on an equitable basis. The stocks offered by us for sale are A1 investments, as we expert properties before listing stocks.

The Pittsburg Mining & Operative Company is under our management and will pay a dividend for the month of June, 1897. This consists of 60 acres of placer, 15 feet deep, going 35 cents a cubic yard; and the Rainy Creek Group, Nos. 1, 2, 3 and 4, on a 50-foot mineral bearing lode or vein, assaying well in gold, copper, silver and nickel.

From Mining in the Pacific Northwest, 1897, by L.K. Hodges for the Seattle P.I., reprinted by Shorey Book Store.

MONTE CRISTO AND ITS MINES

Joe Pearsall's search for mineral deposits in the summer of 1889 led him to the Silver Creek mining district. Good prospects were being found in the vicinity of Mineral City. Pearsall worked his way up the east bank of Silver Creek, then climbed to the top of a mountain from which he could view the terrain to the north. Using an early vintage field glass to survey a distant mountain, he brought into focus the metallic reflection of a huge ore deposit. Emanating from the numerous fissure veins were the telltale stains of oxidized mineral. The mass of ore turned out be galena and iron pyrite. Pearsall, excited about his find, collected a number of ore samples, then returned to civilization to guardedly share his discovery. With the financial backing of John Wilmans who was a mining investor, Pearsall and a friend, Frank Peabody, staked the first claims in the district.

As stories circulated about new and rich discoveries, prospectors with grubstakes traveled to the Monte Cristo mining district. The need for transportation access was apparent from the beginning. The camp, Monte Cristo, lay nestled in a valley at the headwaters of the South Fork of the Sauk River, a difficult place to reach. Prior to the opening of the railroad over Barlow Pass in 1893, a heavily traveled pack trail led over Poodle Dog Pass from Mineral City and Galena to the south - roughly the direction taken by Pearsall when he first entered the district. Another route of travel (wagon road and pack trail) followed along the Sauk River. Around 1900, a visi-

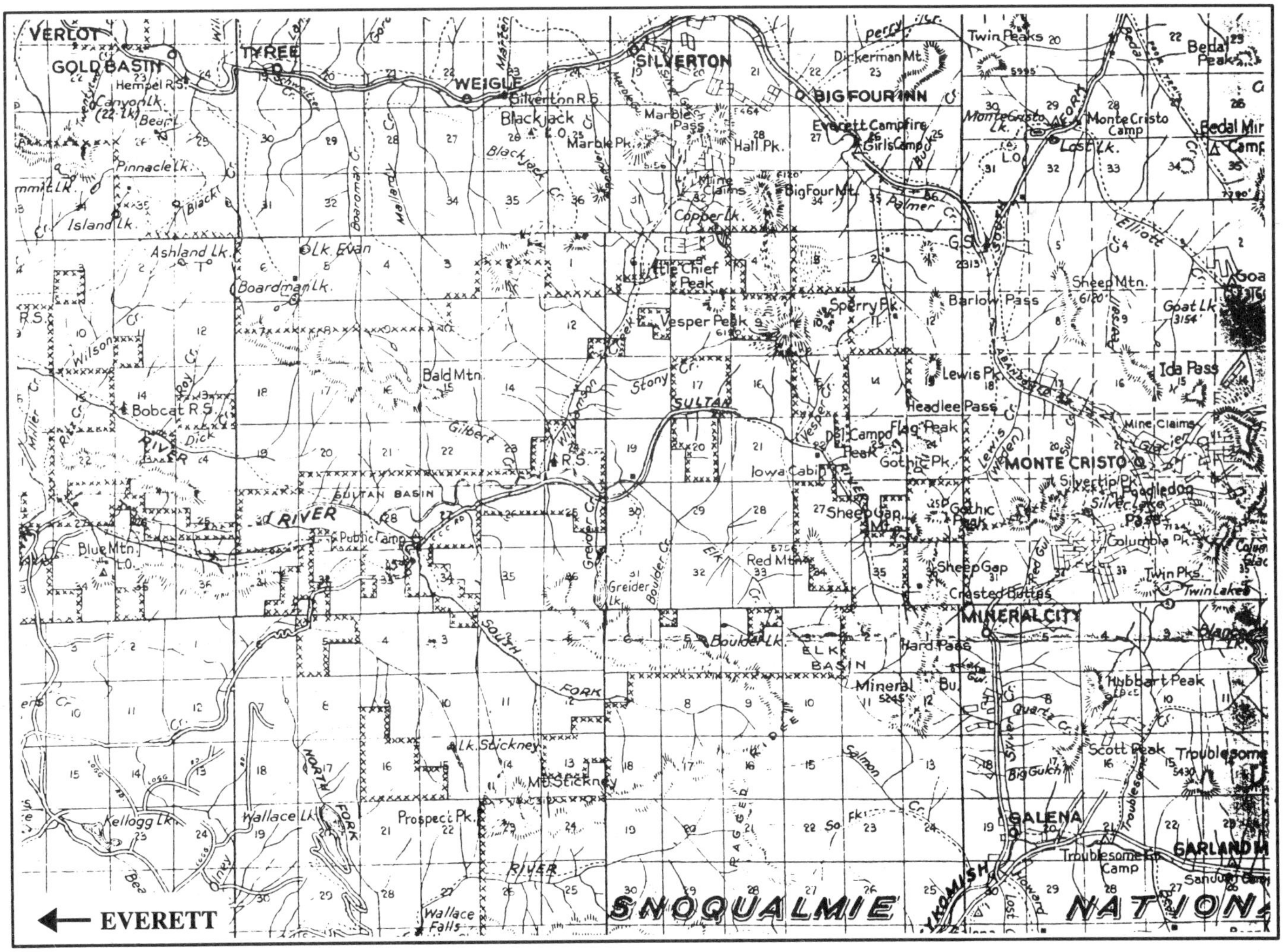

Monte Cristo and vicinity. Used by permission of Metsker's Maps.

tor to Monte Cristo noted the unusual fact that aside from the railroad, only trails were used to reach the town. The Poodle Dog Pass route was still in use.

During its heyday the town supported an estimated population of 1000. Local establishments included a barbershop, railroad depot, post office, mercantile store, hotel with restaurant and bar, two saloons, an assay office, and a one-room schoolhouse.

When the potential for mining investments was made known to wealthy capitalists, several joined in the pursuit to hopefully make millions. More prominent among the list of 19th century investors were the Colby-Hoyt Syndicate and John D. Rockefeller. Large mining operations were soon brought into full production. These were facilitated by the building of a concentrating mill, tramway systems, and a railroad to transport ore concentrates to the smelter in Everett. Buckets suspended on cables conveyed thousands of tons of ore from ledges high on the mountain sides to the concentrator in the valley. As reported by the September 2, 1893, Everett News, the Wilmans Mining Company tramway was now up and running.

> *It runs from the ore bunkers up Wilmans Peak to the Wilmans group of mines, which includes the Monte Cristo, Ouida, Alacante, Hopeful, and Comet, and is nearly a mile in length . . . . At the terminals the cable runs around a sheave, the upper one having two brakes, so that the speed can be governed. The terminal is on wheels and a weight keeps the cable taut. It is run by gravity, the loaded buckets taking up the empty ones. On Monday large timbers were taken up for the ore bunkers at the mines. The tram has a capacity of sixty tons a day and dumps the ore into bins alongside the railroad. The system used is the Haliday (Hallidie), but the Monte Cristo Mining Company will use the Crouch system.*

Monte Cristo. The concentrator at the left is shrouded in smoke. Wilmans Peak is directly above the concentrator and 76 Gulch is to the right. (Photo courtesy of Enid Nordlund)

The Wilmans tram system was reported to have cost about $15,000. As of February, 1894, the tram had deposited 300 tons of ore in its bins to be processed as soon as the concentrator went into operation. The Pride and Mystery trams went into operation in August of 1894. The Pride of the Mountains tramway traveled a distance of about 6000 feet. From the lower tunnel of the Pride, it spanned across Glacier Basin to the ridge of Mystery Hill, then down to the discharge terminal and concentrator near the town.

On December 10, 1894, an accident took place between two towers of the Mystery Tram. The "oiler" was riding a bucket down the tram when the loaded bucket behind him ran away. He had to jump for his life. The drop was from sixty to eighty feet into deep snow. Injuries from the fall were slight, however, it was believed that a piece of ore hit him because a hole was torn in his leg and he suffered several broken bones. He had been warned by the loader not to ride the bucket so the company refused to accept any responsibility for his injuries. [24]

With five levels, the concentrating mill at Monte Cristo was designed to process up to 300 tons of ore per day, 150 tons each side. The mill went into operation in the spring of 1894 and continued off and on until the mines shut down in 1907. During one week in August, 1895, the mill shipped out twelve carloads of concentrates, a record to date. Twelve carloads was the product of about 1,000 tons of ore. A railroad spur ran to the mill where concentrates were loaded onto cars to be transported to the Everett smelter. The mill was reported to have cost about $250,000.

In December, 1895, a snowslide from the roof of the concentrator struck and upset a car of concentrates standing under the eaves ready to be switched out into the down train. The snow was shoveled off and the car replaced with the assistance of blocks, tackles, and a steam tom.

The United Concentration Company's mill building at Monte Cristo. The Mystery and Pride receiving terminal was located to the far left at the end of the long tram shed. At the terminal, the ore was loaded on cars which were pulled by mule through the shed to the concentrator. The Cliff House Hotel can be seen at the upper right. (Herman Siewert photo, courtesy of Enid Nordlund)

The Puget Sound Reduction Company smelter in Everett, Washington, 1893. The complex included a furnace building 95'x50', ore building 114'x112', sampling house 60'x40', roasting house 88'x163', sulphide mill 50'x100', blower house 28'x70', boiler house 39'x50', machine and blacksmith shops, office, and laboratory. There were three blast furnaces, each with a capacity of 80 tons of dressed ore per day or 90 tons per day of raw ore as received at the plant. From January 1 to September 1, 1895, using only one furnace, the smelter produced a total of $546,955.20 in gold and silver bullion, a monthly average of $68,369.40. The plant employed about 150 men. (Photo courtesy of the Everett Public Library. David Ewing photo.)

The Del Campo Mining Company cookhouse and lodging quarters above Weden Lake in the Monte Cristo District. Early mining companys built facilities within walking distance of the mines. It was not uncommon for buildings to be located in such precarious locations. A miners' trail (lower left) follows along the edge of the cliff. (Photo courtesy of Enid Nordlund)

Exhausted miners working long hours ran with a "short fuse." And the fuse was usually burning when it came to meal time, especially if there was a lack of food or the meat was tainted. It was a problem - keeping sufficient meat and other food items on hand to feed a large crew of men. Regular shipments were scheduled but the train was sometimes prevented from arriving on time, delaying the delivery of supplies. In July of 1902, the Monte Cristo Mining Company received two barrels of corned beef, one of which was "rotten," causing the crew to become sick. Another time when there was "trouble in the cookhouse" was in the winter of 1896. The miners from the night shift had come to the eating room and there was not enough food. The cook was run off the hill and a worker assigned his place until another cook could be hired. The Superintendent went up the hill to discuss the problem. He told the miners they should not have taken the matter into their own hands but did admit they had a grievance. [(25)]

The closing of the Monte Cristo mines was partially the result of the numerous business entities which were necessary for the extraction and processing of ore:

> *Mining companies, tramway companies, concentrating companies, and the railroad company all had to make a profit, with the result that instead of one fair mining profit from the operations it was necessary to make several corporations pay dividends. The ore value was not high enough to stand heavy overhead charges, and the result was the ultimate closing down of the principle mines of the camp.* [(26)]

Mystery/Pride cookhouse and eating place in Glacier Basin, In May of 1902, the facility supported a work force of fifty-six men. The weekly meat order was 350 pounds. The X's mark the locations where miners were buried by a snow slide in November of 1898. Sam Strom and a number of co-workers at the Philo mine were caught in the slide. Years later, Strom recalled the catastrophe. He remembered thanking God that he was still alive. One man was killed and four seriously injured, "the rest most scared to death." The Philo mine buildings were destroyed. (Meyer Photo, courtesy of Enid Nordlund)

THE EVERETT AND MONTE CRISTO RAILROAD

In the spring of 1892, construction began on the Everett and Monte Cristo Railroad. The line was completed the following year, allowing travel from Everett to the town of Monte Cristo. During one period an estimated 1500 laborers were employed to build the road. On Sunday, September 17, 1893, the railroad offered the public an excursion run leaving Everett at 7:30 a.m., then departing from Monte Cristo at 3:30 p.m. to arrive back in Everett at 7:30 p.m. Cost for the round trip ticket was $2.50. On September 25th, the company offered a daily run for tourists, excluding Sundays.

An article appeared in the Seattle Post-Intelligencer on September 10, 1893, praising the railroad and mining ventures which justified its construction. The caption was, "MONTE CRISTO CAMP, Things to Convince a Visitor of the Mines' Riches, Great Obstacles Overcome." The thrust of the article was that if anyone should doubt the worth of the Monte Cristo mines, they should travel there by train to judge for themselves:

> *The first impression which will be effectually removed from the doubter's mind will be as to the confidence of those who have built the railroad from Hartford up to the mines. The road is somewhat over forty miles, constructed through a rugged country and under difficulties imposed by nature, well*

The Everett and Monte Cristo Railroad engine no. 3 in the Stillaguamish River canyon. Riprap has been loaded onto a flatcar, probably to be used on the right-of-way. It was bragged that the train moved so smoothly on the track that a passenger could hold a glass full of water without danger of spilling it. With both the concentrator and smelter up and running in the summer of 1894, the company maintained four of the locomotives. The railroad was the only mainline road in the state which had been built solely for mining purposes. The engines, passenger coaches, and other rolling stock were proudly maintained in first class condition. (Photo courtesy of Enid Nordlund)

nigh insurmountable. Yet it will be seen that every part of this road, as it skirts the deep canyon of the Stillaguamish or climbs the mountain side in its winding way, had been built in the most substantial manner. It is not makeshift, but manifestly it has been built for business and permanency. No transcontinental road can in any respect surpass it for solidity and skill of construction. At least a million and three quarters of dollars have been expended on this road.

The article otherwise described in picturesque detail the town, mines and tramways, and surrounding mountainous landscape.

Builders of the railroad could not have anticipated all of the problems which were to be faced in the coming years, often resulting in impassable conditions and the disruption of the delivery of ore concentrate to the smelter. Following the lower elevations of the Stillaguamish Valley and sometimes near the river's edge, the track lay vulnerable to the numerous forces of nature. To be encountered were snow avalanches, rock slides, deep snow accumulations on the track, floodwaters which washed out trestles and undermined the track, and tunnels jammed with debris. Just keeping the track open turned out to be a task of monumental proportions.

Everett & Monte Cristo Railway Co.

WINTER SCHEDULE.

Trains leave Everett Mondays, Wednesdays and Fridays, returning from Monte Cristo same day:

No. East.		No. 2 West.
8:00 a. m.	Lv. Everett Ar.	5:40 p. m
8:10	Lv. Lowell Ar.	5:35
8:50	Lv. Snohomish Ar.	5:40
9:10	Lv. Hartford Jct. Ar.	4:50
9:45	Lv. Granite Falls Ar.	4:30
11:25	Lv. Silverton Ar.	2:40
12:30 p. m.	Ar. Monte Cristo Lv.	1:40

Freight and passenger station at foot of Pacific avenue. General office Wisconsin building.

Everett, Wash.

J. T. McBride, Gen. Manager.

S. N. Baird,

Auditor & Gen. Pass. & Tkt. Agent.

Train schedule, February 12, 1896.

Tunnel No. 4 in the Robe Canyon above Granite Falls is closed by a massive rock slide, a scene repeated numerous times along the track. (Photo courtesy of Enid Nordlund)

The December 12, 1895, Snohomish weekly The Eye, reported the tragic circumstances of a closure of tunnel No. 4.

A slide at tunnel No. 4 prevented the passage of Friday's train, and an extra force of men was sent up to aid the section crew in removing the obstructions. While they were at work, a second slide occurred, killing Patrick Quinn, breaking the leg of a man named Carlson, and injuring foreman Logan and several others. Quinn was a single man, about 25 years old and none of his relatives can be found. After Monday's big double-header train passed up, another slide occurred. This slide was not removed until Tuesday night, in consequence of which passengers from Monte Cristo were transferred to an extra that went up to the slide and returned in the afternoon. The regular train came down in the evening.

SLATE CREEK DISTRICT

In 1895, Charles Ballard was hired by Colonel Thomas Hart, a mining man with financial backing, to build a twelve-mile narrow road from Robinson Creek above Mazama to the Slate Creek District west of the Cascade crest. The summit route previously called Slate Pass then became known as "Hart's Pass." Lode deposits had been discovered in the hills surrounding Slate Creek and miners were rushing into the district to either make wages or seek a prospect.

Two early mining camps in the district were Vera Cruz, located at the headwaters of Bonita Creek, and Bourville, located near the junction of Bonita Creek and Slate Creek. The Snohomish weekly The Eye, on September 5, 1895, announced that the Slate Creek boom had collapsed and that the bustling little mining camp of Vera Cruz was deserted, "its former inhabitants scattering in all directions of the compass." Part of the reason for the announced collapse may have been that fall had arrived in the high country and that prospectors were leaving before bad weather set in. The Eureka and a few other good prospects were continuing development work. Also, shipments by pack train were still being made from Crater Mountain.

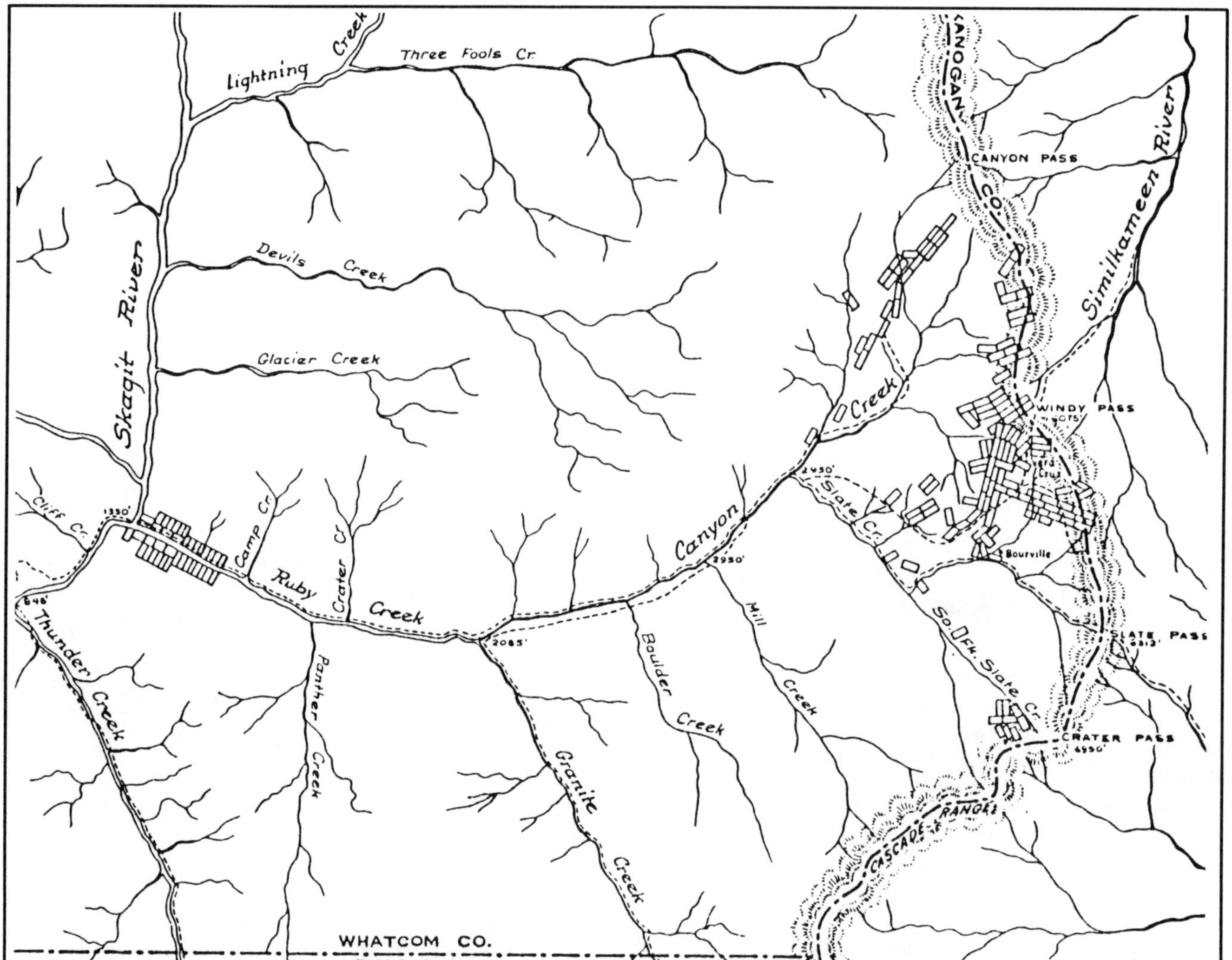

Slate Creek Mining District in 1899. Numerous mines including the Azurite on Mill Creek and the Gold Hill above East Creek had not yet been located. (Reprinted from Moen, 1969, p. 96) [27]

(Above) Scenic view of the Slate Creek District. The New Light mill is at the lower right. Azurite Peak and Mt. Ballard are in the upper left. (Photo from Moen, 1969) [28]

(Below) Alex Barron, on the left, and Guy Waring. Barron prospected up the west slope and discovered the "glory hole." The pioneer merchant, Guy Waring, founded the town of Winthrop in 1891. Waring built a nice log home overlooking the town of Winthrop. The building today houses a portion of the Shafer Museum. (Photo courtesy of Dick Webb)

Near the turn of the century, the outpost Barron became headquarters for miners in the district. The camp was located on Bonita Creek a few miles below Hart's Pass. It was named after Alex Barron who, in 1893, prospected the upper Skagit to a location near the headwaters of Bonita Creek, a tributary of Slate Creek. He was among the hardy breed of prospectors who were now searching the hills for quartz ledges containing gold. Many of the miners had found the placer diggings limited and difficult to work so turned their attention to prospecting for lode veins. And good leads were to be found! Pieces of quartz containing gold had been discovered on Slate Creek. Early accounts record that Alex Barron followed the trail of float (pieces of ore) to its parent source. Here he found the Eureka lode, famous for its glory hole that yielded $120,000 during the first two years of operation. After the initial discovery, prospectors made good wages using a rocker to work the residual placer located on top of the mother lode. When the overburden was finally stripped away, a large quartz ledge was exposed which assayed high in free-milling gold. A miner went into one of the prospect holes at the site and filled his hat with hand-picked samples of quartz showing impressive amounts of native gold. The mine was located at an elevation of around 6,000 feet. During the winter of 1896, workers at the mine were isolated by an accumulation of snow thirty feet deep. At that

Taking supplies into Slate Creek over Hart's Pass was often a challenge. One of Hazard Ballard's horses has gotten mired down in the snow and shovels are being used to dig it free. On the left is Gene Albin, then Harley Wehmeyer with the hat. Hazard Ballard is on the right and Slate Peak is in the distance. The horses are Pete (mired down), Buck, and Ginger. (Photo courtesy of Bill Wehmeyer)

time, as the tunnel was being driven, the ore was changing in character but was still rich with gold.

A method of prospecting similar to that used by Alex Barron was described in The Eye, June 12, 1886. This particular approach had been used for the last year and a half and was called "pocket-hunting." The type of gold sought after was "the pure article" as opposed to that found in quartz or "masses of flinty rock." A pocket-hunter could find a fortune in

Winthrop after the turn of the century. The town was in the path of miners heading into the Slate Creek District. Today, Winthrop is a well-known tourist town with boardwalks and store fronts depicting the theme of the pioneer west.

gold from a barrel-sized hole in the ground. Some deposits had been found with a few ounces while others were reported to have produced thousands of dollars in gold. An eyewitness was once shown a hole a yard-square from which $2,700 had been mined. The deposit was found within a foot of the surface on the hillside (the location was not given). Sale receipts were shown as proof of the amount of gold taken from the prospect. Pocket deposits were said to be found in decomposed quartz, in clay seams, and sometimes in washed gravel. The procedure for finding these deposits was described as follows:

> *. . . The pocket-hunter selects a section where extensive placer mining has been done and where the yield was rich. He conjectures that the gold came from somewhere, and he follows the gulches upstream as far as they have been worked and there takes pans of dirt from the surface and hillsides. If he obtains a "color" or speck of gold from the surface it is a fine prospect, and he follows the trace carefully, taking the next pan full of dirt to be washed from higher ground, and so on until the prospect fails; then he digs for the deposit. Occasionally it is there. Indications are often found where weeks of panning fail to locate from whence they have been washed or thrown; and again, pockets are found by mere accident that have thrown no trace to the surface. . . . A pocket-hunter will carry and wash dirt for days without obtaining a color. When he obtains a speck of gold, however, and if it is the rough, unwashed pocket metal, his chance is fair of finding a deposit, perhaps a fortune. . . .* (At the close of the article, the writer stated his belief that for the amount of labor performed,

Town of Barron. The Methow Trading Company store, run by Guy Waring, is on the right. The narrow-gauge wagon (lower right) was used to transport goods over the narrow Hart's Pass road. (Photo courtesy of the Okanogan County Historical Society.)

both prospectors and pocket-hunters were scantily paid).

Typical of early mining camps, Barron had log cabin saloons (some say seven), dancing girls, a post office, butcher shop, and trading company. Another outpost, Chancellor, was located a short distance to the west where Slate Creek emptied into Canyon Creek. Here were located cabins, a hotel, and a powerplant built in 1905. It has been estimated that during the peak years, upwards of 2,500 miners were camping in the district. The population fell off drastically in 1899 when a large number of prospectors left the hills to join the Klondike gold rush.

THE MAMMOTH MINE

The Mammoth mine was discovered in 1895 by G. C. Matthews and was located on the east side of Bonita Creek, about 3/4 of a mile above its junction with Slate Creek and about a mile southeast of the Eureka lode. Free-milling ore from the mine could be processed by the method of crushing and amalgamation. When Charles Ballard saw ore samples, he was greatly impressed. After further examination, he agreed with a rancher who had grubstaked Matthews to purchase it for the sum of $40,000 in money and stock. This was perhaps in 1897. Ballard had turned his attention to gold mining after silver lost its value and opportunities in the pursuit of that mineral had evaporated.

Having acquired the Mammoth property, Ballard set out to develop its assets. A small steam-powered reduction plant was erected on the corner of the Mammoth claim in 1898. The heavy parts for the five-stamp mill were transported by pack horse. An aerial tram delivered ore several hundred feet from the mine to the mill which crushed the ore to forty mesh. Pete Bryan was in charge of the mining crew. His

Charlie Ballard enjoying a few moments of leisure. The picture may have been taken at the Mammoth mine. (Photo courtesy of Mary Wehmeyer)

brother Matt ran the mill crew. In 1902, under new management, the five-stamp mill was abandoned and a ten-stamp, water-powered mill was installed near the junction of Bonita and Slate Creek. During his management of the mine, Charles sent many shipments of gold by steamboat down the Columbia River. Charles' younger brother, Hazard, transported gold bricks from the mine to Brewster by saddle horse. To protect the cargo, he carried a sawed-off, double-barreled shotgun. One who knew him well was asked if he would have used the gun. His answer was - "He would without batting an eye."

"The way we got our wood, Jan. 19, 1914" The buildings were located at the Mammoth mine in the Slate Creek mining district. The residence on the left is believed to have been the superintendent's home. The person pulling the sled was probably Zora Ballard. (Photo courtesy of Mary Wehmeyer)

Once the miners at the Mammoth ran into a fault (a fracture in the continuity of the rock) and lost the vein. They came out of the mine to inform Charles of the problem. This was the kind of news dreaded by every mine superintendent. When Ballard heard of what had happened, he asked for a broom and a bucket of water. Armed with the unusual items, he went into the mine and proceeded to wash down the face of the drift. When he had finished, a small sliver of a vein was showing. He instructed the miners to follow the lead. They followed it and, as he surmised, were led back to the main body of the vein. Such instinctive maneuvers as this, no doubt, contributed to Charles' reputation as a "mining man."

It was said that some of the ore from the Mammoth was so impregnated with free gold that miners were tempted to high-grade chunks of the rich stuff. Though there is no evidence that they took precautions, Charles and Anna must have been aware of the places where miners hid gold, such as in the cavity of a boot's heel or the false bottom of a lunch bucket. At some mines, workers were required to change clothes after a work shift to discourage the high-grading of gold.

A friend of Anna's who attended the Priscilla Club with her has shared Anna's story about the time when she first met Charles. Miss Anna Elliott had come to the Okanogan around 1888. As the story goes, she was working as a maid at a hotel in Ruby. (The hotel may have been in Conconully. It's even possible that she was a relative of the owners of the Elliott

Anna Ballard with relatives. The lady on the left is probably Mrs. Hank Johnson, Zora Ballard's mother. Next to her is Anna, then Zora and Hazard Ballard. (Courtesy of Bill Wehmeyer)

Hotel in Conconully. The hotel was the largest building in town, 40'x70' feet. It is known that Anna and Charles courted at Conconully.) Charles came and sat down on the porch. As he was sitting there, the young lady sweeping the floor caught his attention. The energy with which she performed her task greatly impressed him. With affection, Anna remembered that he was taken with her especially because, in her words - "I think he liked the way I stepped about." This marked the beginning of a long, loving, and adventuresome relationship. They were married in 1893. Over the next four decades, Anna stayed loyal to Charles even during the hardest of times. Once things got so lean that they had to shoot rabbits and eat berries to survive. She supported him in his wilderness ventures - blazing trails through the underbrush, setting up camps, and packing grubstakes.

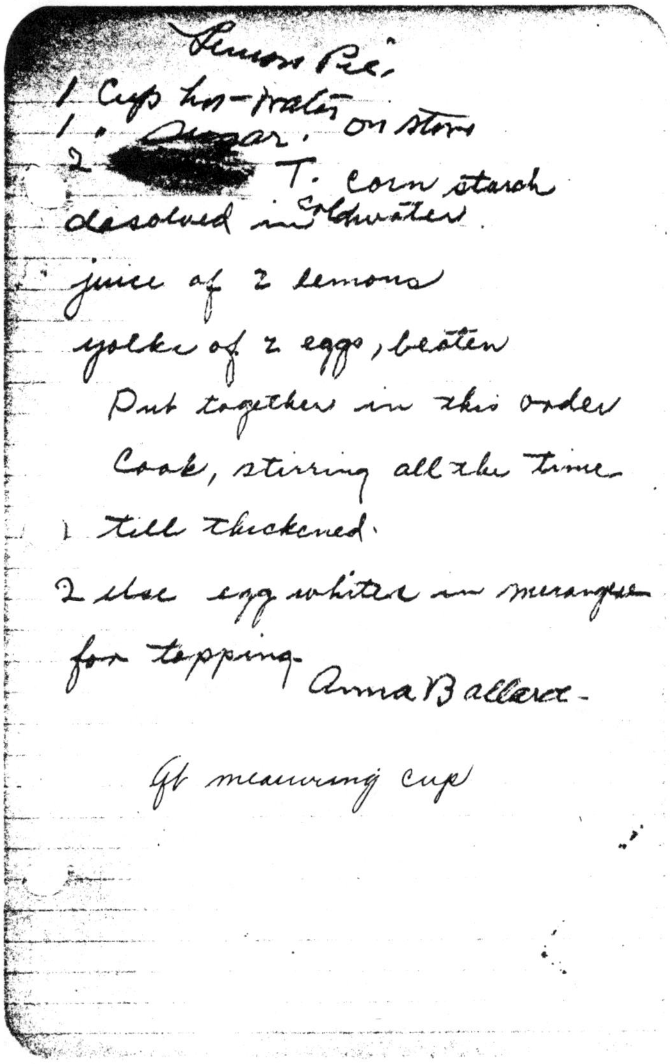
Lemon Pie
1 Cups hot-water
1 " sugar. on stove
2 T. corn starch
dissolved in coldwater.
juice of 2 lemons
yolks of 2 eggs, beaten
Put together in this order
Cook, stirring all the time
till thickened.
2 use egg whites in meringue
for topping. Anna Ballard -
Qt measuring cup

Lemon pie recipe written by Anna Ballard and given in 1956 to Cathy McCauley for her bridal shower cookbook. Could Anna's good cooking as reflected in this recipe have been one of the hidden ingredients to Charles' unending drive for success?

Mrs. Charles (Anna) Ballard in an interview by Hu Blonk, published in the Wenatchee Daily World, Sunday, April 10, 1960, recalled that a crew worked the first summer and winter at the Mammoth before finding rich ore in the spring. During one period, forty men were employed at the mine which produced good returns. The Ballards had once stacked $60,000 worth of ore in their cabin, which was later moved to the cellar. Anna remembered that life was difficult in the high country:

> *"The first winter (1899) there was an awful fall of snow," she recalled. They couldn't get in or out from November until spring, except on snowshoes and that was hazardous below the steep snow slopes. Life was difficult in the high country. A woodstove and coal oil lamp had to suffice. A pack train brought in goods a couple times a week. There was no social life, only hard drinking by the miners in Barron, a settlement scattered of houses and buildings from which Mrs. Ballard stayed away. . . .One time they snowshoed out in the dead of winter, making only half the 12-mile walk from the Mammoth Mine to Robinson Creek Village by night fall. The Ballards crawled under a shed, unrolled their blankets and slept through the frigid night.* [29]

The Mammoth ceased operations in 1901. It was probably right after the shutdown that Charles ran a placer operation on Slate Creek below the Mammoth millsite. The owner of a collector-sized nugget believes it to have come from the placer diggings. In 1904, Charles and Anna traveled to the gold regions of Arizona and Nevada where they spent the next two years, then returned to Washington State.

Vernon and Beulah LaMotte, long-time residents of the Methow Valley. Vernon is a retired mining engineer and knowledgeable historian of early mining. In 1934 and 1935, LaMotte was employed as an assayer and shift boss at the Republic mine. In 1936, he returned to the Methow where he worked in the Red Shirt and Alder mines. During the war, he worked for the U.S. Bureau of Mines doing exploratory work in search of critical war minerals.

Vernon LaMotte, retired mining engineer, recalls numerous stories from the mines of the Methow and Slate Creek District. Some of these stories did not have a happy ending. Mining was a dangerous business. At noon one day, holes were being blasted for the footing of the headframe which was being installed above the shaft at the Mammoth mine. To alert the camp that a charge was going to be set off, someone yelled "fire." The cook came out of the cookhouse to watch the rocks fly. To his misfortune, a rock came down and struck him on the head. He was fatally injured. Such accidents were all too common at the mines. A man is buried near the trail below the Minnesota mine. (The mine was located by Bill Voight in 1904). He stopped off at Voights on his way up Canyon Creek to the town of Barron. A snowslide had closed off the portal of the mine so the man was hired to shovel it out. While digging in the snow, another slide came down the narrow canyon and killed him.

Another cause of injury at the mines was the miscalculation of the time that fuse was to burn before setting off dynamite. Miners could save time and money by trimming the fuse at a marginally safe length. A miner is said to have been buried on the hillside near the Gold Hill mine. It is uncertain as to what really happened. He may have cut the fuse too short or the fuse may have "jumped." Sometimes, if a fuse had a bad section, (possibly caused from the powder being packed too loosely) that portion would burn faster than it was intended to burn. This was known among miners as "jumping."

Tools of a mining engineer: pocket transit; Robert Peele's Mining Engineers' Handbook Vol. 1; and field book used by Vernon LaMotte during his years as a mining engineer. The transit was used for tracing the path of a vein and to determine angles and directions.

THE AZURITE MINE

The Azurite, Mammoth, and numerous other mines were located in the Slate Creek District northwest of Mazama, Washington. Though the mines were located in canyons which drain to the west, major access to the mines was from the east out of the Methow Valley. The Azurite was located on the northeast slope of Majestic Mountain in the Mill Creek canyon, about 3½ miles below Azurite Pass.

According to Mrs. Charles (Anna) Ballard, Hazard Ballard discovered the Azurite lode while he was running a pack string into the country. It is known that he packed groceries and supplies from Barron down to Ruby Creek so would have been familiar with the mountains and valleys of the district. During one of his trips down or up Canyon Creek, he may have ventured up Mill Creek to prospect or take a shortcut over Azurite Pass, thus could have discovered the Azurite ore deposit. Apparently prospectors had already explored the area as evidenced by an old cabin located near the mine. The Ballards referred to it as the "dinky cabin." The Azurite mine was located about five miles from the mouth of Mill Creek.

Other sources point to C. R. (Dick) McLean and Pete Bryan as also having made early discoveries at the Azurite. McLean ran a pack train from the Methow to the Slate Creek District in the 1890's. In 1915, the original claims of the Azurite group were filed by Charles Ballard, his brother, Hazard, and C.R. McLean.

In 1918, the Discovery Tunnel was run. It may have been the following year when Charles, Anna, Hazard and young nephew, Donald Ballard, rode by horseback to the mine. During their trip, they stayed overnight at the bunkhouse where Hazard and his father-in-law, Hank Johnson, had a sawmill. This was about six miles above Winthrop on the Methow River. More than anything else, Don remembered the place being infested with mosquitos. Logs were brought into a millpond and slough area which provided good breeding ground for the creatures. He got more than his share of bites. They rode over Hart's Pass to the Mammoth where they bunked down for the night, then rode on over Cady Pass to the Seattle camp or "lower camp" where there was an abandoned two-story building. It had served as the cookhouse and bunkhouse for the Seattle mine. The mine was located on the west side of Mill Creek across from the building, about 2¼ miles below the Azurite. While Anna got things ready for dinner, Hazard went out to catch some fish. He had no problem harvesting more than they needed, for in a short time he had caught 135 trout. Don recalled that they did not lack for food that evening. That night, they stayed upstairs in the two-story building. When Hazard returned with his horses to Lost River, he invited Don to go with him but the young man decided to stay with Charles and Anna until they came out in the fall. During the following weeks, they didn't do much mining but spent a lot of time building trails and doing other surface projects. At that time, the Discovery Tunnel had been driven in about fifty feet to a vertical shaft of twenty feet.

Over the next several years a small camp was built near the mine and development work was continued along the Azurite vein. Some of the assays ran high, into the hundreds of dollars per ton. In the hopes of producing some tangible evidence of the mine's potential, Charles designed and built an overshot, water-powered arrastra using sheet iron, stones, cement, chains, and wooden pins. Hazard used his pack animals to haul materials for the arrastra as far as the Seattle camp on the east bank of Mill Creek. Charles was unable to persuade his brother to move the supplies any closer to the mine. When finished, the arrastra was equipped with two tanks made from sheet iron over a base of large stones. They were separated about two or three feet making space for a vertical overshot wheel which powered two flat grinding wheels, one for each basin. Wooden pegs, near the center of the water wheel, engaged pins set near the perimeter of the flat wheels making them turn. As they turned, the flat wheels pulled millstones (boulders) over chunks of ore. Ten tons of average ore was packed from the mine to the mill. Building the plant, mining the ore, breaking it into small pieces, hauling it to the mill, then feeding it into the trough had to have been backbreaking work.

In addition to designing the mill, Charles used his engineering skills to devise a system using cyanide to extract the gold. Mining engineer, Vernon LaMotte, visited the site in 1932 and found the leach tank still in usable condition. The tank, about four feet tall and six to eight feet across, had been made from pieces of sheet iron riveted together to form a rim over a cement floor. The floor sloped toward a drain located at the center. Slats about 1" x 1" were placed on the bottom (as spokes of a wheel), running from the outside to the drain. Crushed ore was shoveled onto a burlap mat which rested on top of the slats. As the cyanide filtered through the crushed ore, it dissolved the gold. When the enriched value-bearing solution reached the bottom, it passed through the mat, then flowed between the slats to the drain. From the drain, it ran through a small pipe to zinc boxes where zinc shavings precipitated the gold out of the value-bearing solution. The black sludge product was melted in a furnace to recover the gold.

The brothers must have been delighted when they saw the results of their effort - a 19½ ounce bar of gold. Charles was interested in having a good sample of the precious metal to show when promoting the mine. The brick would serve that purpose well. And it would serve another purpose, too. It

NOTE: This Certificate is for Series A, Fully paid and Non-Assessable Stock.

INCORPORATED UNDER THE LAWS OF THE STATE OF WASHINGTON

Series A No 130 Shares 1500

AZURITE GOLD COMPANY

CAPITAL STOCK $30,000: 3,000,000 SHARES, PAR VALUE ONE CENT

Series A: 1,450,000 Shares, Fully Paid and Non-Assessable
Series B: 1,550,000 Shares, Fully Paid But Assessable

This Certifies that H. E. Hinds is the owner of One thousand five hundred Shares of the Capital Stock of AZURITE GOLD COMPANY transferable only on the books of the Corporation by the holder hereof in person or by Attorney upon surrender of this Certificate properly endorsed.

In Witness Whereof, the said Corporation has caused this Certificate to be signed by its duly authorized officers and to be sealed with the Seal of the Corporation this 13th day of February A.D. 1926

Arthur C. Ballard, Secretary — Chas. H. Ballard, President

Azurite Gold Company Corporate Seal 1925

SHARES ONE CENT EACH

©GOES 23

Azurite Mine certificate. On February 15, 1926, H. E. Hinds purchased 1500 shares of Azurite stock. Mr. Hinds and his wife cooked at the Azurite for Charles. (Courtesy of Vernon LaMotte)

provided Charles with some amusement - that is, observing the reactions of those seeing his huge chunk of gold. Dr. Donald Ballard, nephew of Charles, recalled an incident which involved the famous gold bar. Charles was at the home of his brother, Arthur, in Auburn. At dinner time, he was showing off the heavy piece when it slipped from his hand and broke a fine Haviland China teacup belonging to Donald's mother. Because of the value of the gold and what it meant to Charles, she was quite willing to forgive him.

While traveling, Charles had to give thought as to how to protect the gold brick. There was always the possibility of being robbed or somehow just losing the thing. He had the reputation of being a mining man and it may have been rumored that he was carrying gold.

Needing a place to hide the gold while staying at a hotel, Charles came up with the idea of putting the bar in the toilet tank. (He usually stayed at the Arlington Hotel during layovers in Seattle and probably was staying there at the time). Who would ever think of looking in the tank of a toilet for something valuable? He carefully tied a string around the brick (about 2"x3"x¾ ") and lowered it into the tank not realizing that it would interfere with the flushing mechanism. He left the gold in its hiding place and went about his business. While he was away, the hotel maid went into his room to make the bed and heard water running in the stool. She went to check it out and found the heavy metal piece. She probably recognized it as gold, for nothing else that size could have weighed that much. The occupant was nowhere around so she decided it would be best to give it to the desk clerk. When Ballard returned to the hotel, the clerk asked him about the object that had been found in the tank. He was very surprised to see the gold. And to think that it had been passed around! He was glad to get it back and grateful for the honesty of the maid and the clerk.

About 100 feet below the Discovery Tunnel, the Burnam Tunnel was driven seventy feet along the Azurite vein. In 1925, the Azurite Copper Company of Delaware was reorganized as the Azurite Gold Company of Twisp, Washington, with Charles Ballard as president and superintendent. When the mine was first discovered, it was thought that the main value was in copper. Azurite is a blue ore of copper mineral. However, it became apparent that the greater value was in gold so the company was reincorporated, keeping the name Azurite but substituting the word "gold" for the word copper. Because of the wilderness location of the mine and narrow trail access, improvements were costly in both time and labor. In October of 1929, an air compressor, machine drills, gasoline in cans, powder, and other supplies were dragged or carried to the mine by horse at the cost of $100 per ton.

Beginning in April of 1930, a road was con-

Robinson Creek supply depot, sometimes called "the end of the road," located about nine miles above Mazama. Dick Horn and co-workers are loading supplies for a trip to the Azurite. At that time, 1930, four Caterpillar tractors with trailers and two Fords were used to haul supplies. Robinson Creek was the last place miners visited before heading over Hart's Pass. Here was a hotel, a store run by Guy Waring, blacksmith shop, and saloon. (Azurite Prospectus)

The "puddlejumper" was a modified Model T Ford used to travel to the Azurite mine during the summer months. It was equipped with an extra transmission, extra water pump, extra oil line to the engine, semi-balloon tires on the rear, and narrow wheel axels to accommodate the narrow gauge road over Hart's and Cady Passes. Charles Ballard's nephew, Donald, is in the drivers seat. The two passengers are unidentified. Ballard had offered to pay $1,000 if George (Buck) Therriault of Twisp could build a car to get him into the mine and out the same day. When the car was made ready, they left the Robinson Creek cabins at 7:00 a.m. and arrived at the mine at dinner time. They left the Azurite at 3:00 p.m. and arrived back at Robinson Creek at 4:55 p.m. Ballard was impressed with the performance of the car and called it quite a "puddlejumper." Being a man of his word, he paid the $1,000. Donald later became Doctor Ballard, PhD., M.D. When he was interviewed in 1994, Dr. Ballard still vividly remembered driving the puddlejumper for Charles. He said it was a "fun" car. With the three-speed transmission, Ruckstell axel, and original Ford transmission, it was capable of about twelve forward speeds from extra low to super overdrive. (Photo from the Horn-Olling collection, Shafer Museum)

structed from Lost River to the mine, a distance of twenty-six miles. The last stretch over Cady Mountain was difficult to grade because of the bedrock which lay near the surface. In spite of pressure to hurry its completion, the road was built at no more than a 10% grade. Maintaining an easy grade necessitated a number of switchbacks on Cady Mountain. There were at least six on the descent into Mill Creek canyon. On November 1st, Charles, with his wife, Anna, traveled in a Model T Ford from Robinson Creek to the portal of the Tinson Tunnel in only five hours. Formerly, by horseback, the trip would have been at least a hard day's ride. Anna was a good passenger and didn't complain about the rough ride. The narrow road provided vantage points of spectacular scenery, however, it was not for the average sight-seer. A traveler who later went over the road spoke of it as being "a wild and dizzy ride."

At the turn of the century, miners with horses could expect an adventuresome climb from Robinson Creek to the summit. The most hair-raising part of the trail was called "Deadhorse

Point." To get their horses around a sheer cliff that dropped several hundred feet, a precarious catwalk had been built using planks laid on steel rods. The rods were anchored in holes which had been bored into the rock. The place received its notorious name because of the time when a string of horses tied together went over the edge. One horse fell, pulling the others off the narrow walkway.

After the road was completed in 1930, a No. 15 Caterpillar tractor could haul up to two tons to the mine in a single day. Whereas the cost for transportation by horse had been $100 or more per ton, now the cost was reduced to about $20 per ton. Food, supplies, and mining equipment had to be delivered to the mine in the fall before the long winter lock-up. Quantities had to last until the end of June. The road over Hart's Pass (6,197 feet) and Cady Pass (6,000 feet) was usually blocked by snow from the middle of October until the middle of the following June. One miner referred to the long winter months at the Azurite as "Nine months rest from noise and excitement." Another spoke of it as "Nine months of imprisonment." And it is understandable that because of the isolation during the winter months, it was a "no woman" camp.

The Azurite miners were accustomed to working and living in close quarters, both at the mine and in the bunkhouse. The bunkhouse had double-deck bunks and not a lot of extra room. Between the two rooms of the bunkhouse was a recreation room with a stove. Inside activities included weekly movies, pool, playing cards, and reading. In spite of being crowded together, the men got along quite well.

The bunkhouse covered with snow. (Horn-Olling collection)

A worker, Don Olling, recalled that the snow was so deep during the winter that no daylight could be seen from inside the bunkhouse for five months. Snow depths of twenty feet were not uncommon. The only visible evidence of the building from the outside was the emission of smoke. The entrance to the living quarters was from the roof, about twenty to thirty steps down to the main floor. To protect the building from the frequent snowslides which came down the mountain, it had been built into the hillside. Miners, safe and secure in their bunks, could sometimes hear the rumble of the snow as it cascaded down the mountainside. A miner was once thrown down the stairs by a snowslide. There was an open area between the bunkhouse and the snowshed which protected the large diesel fuel tanks. Before crossing the area, miners always hesitated to detect any movement of air or sound which might indicate a descending slide. As the miner approached the building, he felt a telltale rush of air so knew he'd better move fast. He made a mad dash for the entry and barely made it when a deluge of snow hit him on the backside, catapulting him down the stairs. Don Olling was in another room when he heard the commotion. There was the awfullest swearing. He went to investigate and discovered a pile of snow at the bottom of the stairs and one very distraught miner.

At times, up to eighty men slept in the bunkhouse. While the men normally kept themselves neat and clean, some of them had a real problem with sweaty feet. Many years after living in the bunkhouse, one of the miners still remembered the repugnant smell of stinking feet. He exclaimed, "some of those men would let themselves get rotten!" Then there was the pervasive wake-up call in the early morning when the bull cook (assistant cook) rousted out the miners. The rhetoric was almost as imposing as the stinking feet. The cook broadcast in rhyme - "Roll out, roll out, not one, not two, but the whole d_ _ _ crew."

One time the bull cook, Charlie Graves, went out to the spring to get some water. A trench had been shoveled through the deep snow to the spring. He had just filled the buckets and stood up when a "dry" slide came down and buried him under several feet of snow. A dry slide was termed as such because the surface snow was dry and avalanched on top of the snowpack. A wet slide was usually in the spring or during a chinook when the melting snowpack itself let loose, down to the dirt level. The latter was more devastating, capable of taking out trees, buildings, or whatever else might be in its path. In a short time the cook realized that Graves was missing. Dick Horn and others decided to search beneath the snow at the water hole. During the search, Horn probed into the snow and got lucky. He hit Graves' metal hat. It took several minutes for the rescue party to dig him out. To Graves, it seemed like he'd been in there an hour. Released from his prison in the snow, he spat profusely from the chew of tobacco he still had in his mouth. Under all that snow he had no place to spit. Visibly shaken, he retreated to his bunk to recuperate from the ordeal. He was back on the job the next day.

During the fall and winter of 1929, the Tinson Tunnel was driven several hundred feet with a showing of one to three feet of rich sulphide ore. At a depth of more than 500 feet, the ore carried fine particles of free gold that could be separated by panning. This tunnel was named after William Tinson, a New York laundry owner who had assisted Charles with financing. In 1917, Ballard had gone to New York and lined up several capitalists to back his venture. He was about to complete a deal when President Wilson declared war on Germany. Probably because of the war, all of the potential investors backed out except Tinson. It was most likely Tinson's financing that enabled the running of the Tinson Tunnel which blocked out a huge amount of high-grade ore. While Arthur Ballard was secretary of the Azurite Gold Company, Wm. Tinson served as assistant secretary.

From the Tinson Tunnel, the Cooper raise was cut in ore 275 feet to the surface. Encouraged by the initial results of this work, Ballard started the Wenatchee Tunnel during the winter of 1930-31. The portal of the Wenatchee was located below the Tinson Tunnel and near the camp at an elevation of 4,875 feet. Charles believed this cross-cut tunnel

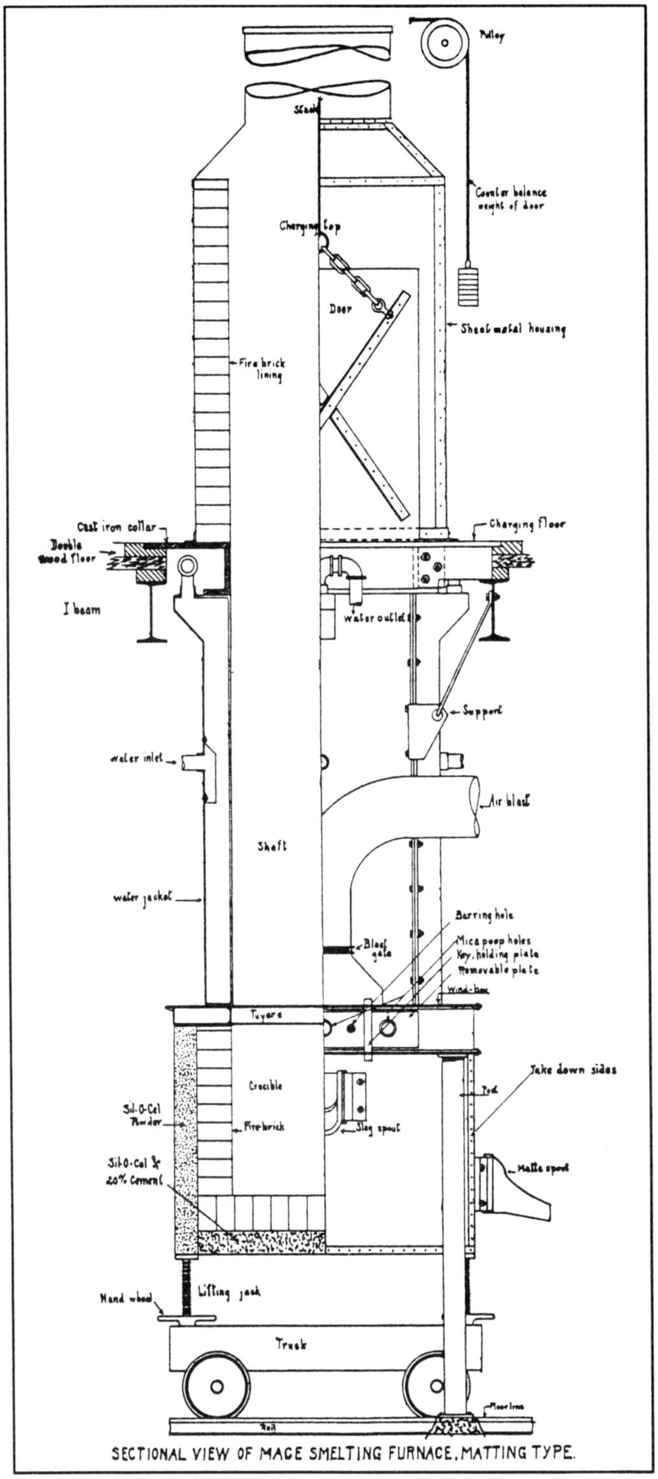

The Mace Smelting Furnace illustrated in an advertisement brochure. A smelter was a blast furnace which reduced metals from their ores by a process that included fusion. The precious metals were collected in a carrier called matte. The waste product was slag.

would intersect the Azurite vein at around 1,300 feet from the portal and about 500 feet vertically below the Tinson Tunnel.

In March of 1930, a contract was drawn up for the purchase of a 30-ton reduction plant from the Mace Smelter Company of Denver, Colorado. As of the winter of 1930-31, most of the parts for the smelter were on location at the mine. The plant was up and running the following spring. Ore samples from the mine were frequently analyzed to determine their value. Assays provided information necessary to avoid the mining and treatment of unprofitable material. During one summer, Donald Ballard assisted his uncle Charles with test analysis in the assay office. Donald had graduated from the University of Washington with a major in chemistry and was eager to apply his knowledge. Also, working in the assay house provided him the opportunity of working under Charles' supervision. His uncle was an acknowledged expert in the field of assaying. The equipment, too, was of top quality. Charles was said to have owned an assay balance of such precision

Dr. Donald Ballard at his home in 1994, displaying gold nuggets and gold brick.

100-ton cyanide plant above Mill Creek, opposite the Wenatchee tunnel and blacksmith shop. The cable-suspended footbridge leads to a covered stairway with the assay office shown below the mill. The large barrels contained fuel to run the diesel electric plant. Tramway ore buckets were used to transport ore across the canyon to the mill. Each of the four buckets carried 400 pounds of ore. (Horn-Olling Collection)

that it registered the weight of a pencil mark. While working at the mine, Donald melted together all of the gold from the assay beads and several decades later, in 1994, still had in his possession a miniature gold brick from the workings. The smallness of the brick is understandable considering the size of each bead. If the size of the gold and silver bead were as large as a common pinhead, the sample would run several hundred dollars a ton at the old rate of $20.67 a Troy ounce. To determine the amount of gold, the silver had to be dissolved out with nitric acid and the remaining gold weighed.

Donald and his uncle shared a special friendship. Each year they celebrated their birthdays on the same day - August 26th. Donald was born in 1908, and Charles fifty years earlier in 1858.

Metallurgists believed that the ore was of the type compatible with the smelting process. When the Mace Smelter went into operation, however, it did not prove as successful as had been predicted. Ballard was one to "take the bull by the horns," so he persuaded Mr. Mace to come to the Azurite to adjust the operation and hopefully improve the concentration of matte. The problem was apparently one of design. According to one explanation, the plant wasn't tall enough to give enough time for the iron pyrite to oxidize to iron oxide and thus reduce the weight and volume of the matte. To have the heavy matte ore product refined, it had to be taken out over a rough and narrow road system, then transported down the Methow Valley to Pateros. From there it was taken all the way to the smelter in Tacoma. It was ninety miles from the mine to the rail station at Pateros. A shipment of fifteen tons in 1933 yielded $2,344 in gold, $29 in copper, and $8 in silver. Charles Ballard had always believed that the most economical treatment of the ores was by the use of cyanide. He was to be proven correct.

In the fall of 1933, L. A. Lavensaler, a consulting engineer, did an inspection of the Azurite mine for the American Smelting and Refining Company (ASARCO). The findings were favorable. In January of 1934, Charles and a representative of ASARCO negotiated an agreement in which the company would lease thirty-six claims and six mill sites for a period of twenty-five years. The Azurite Gold Company would be paid 50% of the net smelter returns. When the road became passable in June of 1934, more engineers visited the mine and in July, development work began in earnest.

The Wenatchee Tunnel had been driven eighty-seven feet when ASARCO took over in January, 1934. In 1935, the Wenatchee Tunnel was driven 1,308 feet before reaching the Azurite vein, close to the projected distance. A drift was then driven a total of 1,275 feet along the vein and a raise cut 627 feet to the Tinson Tunnel, with stations at 250 and 450 feet. During the early part of 1936, ASARCO made the decision to launch into full production. A 100-ton capacity cyanide ore-dressing plant, tramways, and other facilities were constructed on the property.

Charles died at the close of 1934, never having the opportunity of seeing his dreams for the mine fulfilled. While attending a mining congress in Spokane, he became ill. Having pneumonia, his condition worsened while on the train during his return trip home. He was taken to St. Anthony's Hospital in Wenatchee where on December 29, 1934, he passed away at the age of 76. He was laid to rest at the Washelli Cemetery in Seattle. Overlooking the wilderness country in which he spent nearly thirty years stands a lofty mountain, Mount Ballard, an enduring tribute to the Ballard name. After Charles' death, his wife, Anna, was elected to take his place as president of the Azurite Gold Company. The mill operated until February, 1939, and produced $972,000 in gold and silver.

The May, 1938, Washington Prospector - Miner, presented the 1937 annual report of the Azurite Mine. During the year, 27,000 tons of ore were mined and milled, yielding 12,687 ounces gold and 1,108 ounces silver. Sales of metal during 1937 were 11,746 ounces gold and 972 ounces of silver, which brought a return of $406,632 above the cost of freight and treatment and yielded an operating profit of $131,308. The number of employees averaged sixty-three.

WORK DUTIES AT THE AZURITE
DON OLLING

Don Olling and co-author Vida.

Don Olling worked at the Azurite four winters and five summers, from 1934 to 1938. During these years he performed a number of important duties. In the summer he worked at the mine and drove bulldozers which were used to haul supplies and keep the road open to the mine. The small "15" and "22" cats were ideal for traveling the narrow-gauge road. When hauling supplies, the drivers had to back up several times to negotiate the switchbacks on Cady Mountain. Special grouser cleats were installed on the tracks for traction in the snow and ice. Because of heavy snowfall, the road was only passable during the summer months.

In addition to the cats, modified vehicles with a narrow wheelbase were used to haul supplies to the mine. In the spring of 1936, George and Sumner (Stony) Stonebraker from Idaho contracted to haul freight to the mine using trucks fashioned from International school bus chassis. These were cut down to fit the narrow road. It was probably in 1936 when Olling and other workers widened the road an additional two feet from Hart's Pass to the mine on Mill Creek. Normal width vehicles were then able to travel the whole distance to the mine, although the narrow vehicles were apparently still used.

During the winter months, Olling ran the generators in the powerhouse. They provided electricity for mining equipment and compressors which ran the pneumatic drills and were powered by two Chicago diesel engines. He described the starting of the engines:

> *At one end of them you had a big wheel similar to a helm wheel on a boat. It was a wheel with handholds every so often around it. And then it had different signs around on it that would say COMPRESSION OFF,*

Inside the generator plant. There were two 400 horsepower Chicago pneumatic diesels on generators. (Horn-Olling Collection)

COMPRESSION ON, or RUN. When you got all ready, you turned the air valve on and the compressed air and you threw the motor so the compression was off. That compressed air would spin the motor, and after it spun it up to the correct speed you'd throw this wheel into COMPRESSION and then RUN. When it went, it went into the RUN position, the diesel would start feeding, and when it went in through the injectors it would fire and run. [(30)]

Olling was a capable and trusted individual. Soon after the cyanide plant went into operation, he was quietly awakened early (3:30 a.m.) in the morning by the superintendent of the mill, Mr. Coombe. He was told to get up, go to the kitchen for breakfast, then report to the mine office. Upon arriving at the office across the canyon, the superintendent informed him that they were going to deliver a gold shipment to Pateros. He was given the choice of carrying a sawed-off shotgun or .45 caliber Colt. He chose the Colt. Each gold brick, about the size of an ordinary clay brick and worth about $40,000, was carefully removed from the safe and carried to the vehicles where a blanket was thrown over them. Olling was caught off guard when he attempted to pick up the first brick. The superintendent chuckled as Don struggled to handle the brick. He later said, "I felt silly staggering out to the truck with that little brick in my hands!" The bricks were split between the superintendent who drove the lead vehicle and Don who followed at a safe distance in a Model A Ford with a flatbed. They drove nonstop to their destination which was the railroad depot at Pateros. When they arrived, Wells Fargo guards armed with machine guns were there waiting to receive the shipment.

In December of 1936, a snowslide took out the ore bin and damaged the aerial tram going from the bin to the mill. New cable was needed to repair the tram. The road was closed to vehicles because of snow so the only possible means of delivering the cable to the mine was by air. Johnson Flying Service in Cascade, Idaho, owned a Ford tri-motor plane with a door opening large enough to load the spools on edge. Two flights were made from the Twisp-Winthrop air strip, one for each 1,600-pound spool. Don Olling went along as navigator for the first flight. A block of wood was wedged under each side of the heavy spool to keep it in place. As they approached the drop site, the plane was tilted one way so the front block could be removed, then the other direction so the spool could roll out of the side door. The spool from the second flight barely missed hitting the catwalk which spanned across Mill Creek. Considering the weight of the cargo and location of the mine in the narrow valley, the emergency air-drops were a remarkable accomplishment.

The Ford tri-motor plane which flew into Mill Creek canyon to deliver two spools of cable. This picture was taken just after one of the spools was dropped. (Horn-Olling collection)

THE KIKENDALL BROTHERS AND THEIR DOG TEAMS

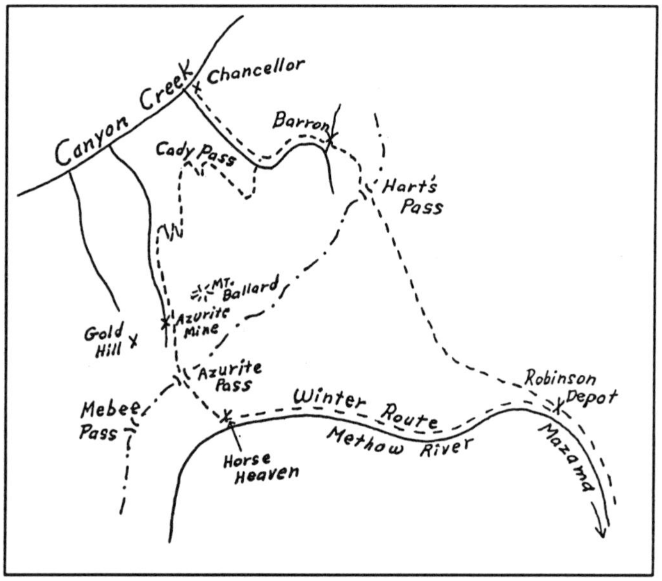

During the summer months travel to the Azurite was over Hart's Pass to Slate Creek, then over Cady Pass to the mine on Mill Creek. Traveling the switchbacks on the descent from Cady Pass to the Azurite Mine was a great challenge, sometimes requiring haulers to back up several times to negotiate sharp turns. During the winter months travel was by dogsled or snowshoe and was usually over the steeper but shorter Azurite Pass route. The distance to the mine from trail's end at Robinson Creek was around twenty miles. The company maintained a building at Robinson Creek.

Chuck and Ed "Kike" Kikendall had been hired by the Stonebraker brothers from Idaho to transport mail and supplies to the mine on Mill Creek. The contract required four trips per month and a minimum of eighty pounds each way. They were paid $6 a trip and an additional 20¢ per pound for weight exceeding eighty pounds. A round trip took two days or longer depending on the weather. The dogs were harnessed two by two with a lead dog in front. Ed drove the heavier sled so usually mushed with seven dogs while Chuck ran five. The size of the team was determined by the size of the load.

Once, during the late spring, Ed used a twelve-dog team with his lead dog Dewey to haul $2,000 worth of machinist tools from the mine over Cady and Hart's Passes. As they rounded Dead Horse Point, a mountain goat and her kid jumped onto the trail. The dogs gave pursuit and in doing so nearly pulled the sled over the cliff. Ed's commanding whistle saved the day. Dewey responded just in time to keep the team from plunging hundreds of feet to the valley below. The sled came within inches of

Mushers, Ed and Chuck Kikendall, with their dogs during the winter of 1937-38. The men had stopped for a rest at the summit of Azurite Pass. The dogs were let loose at this point and the sleds steered down a steep quarter-mile grade. On this trip, Ed (on the right) used his lead dog, "King," a white Siberian Husky shown lying down near the front of the pack. With intimidating blue eyes, he was the leader but a real gentleman. Ed's wheel dogs were a Lab-Husky named "Stub," and a German Shepherd called "Tim." A strong wheel team (immediately ahead of the sled) was necessary to complete the pull around a sharp turn or switchback. The dog standing at the front is Chuck's lead dog, "Cap," a black Pinscher. (Horn-Olling Collection)

Chuck Kikendall with his dog team. Notice the 5 stamps in the background. These appear to be the stamps which had been used at the steam-driven mill installed at the Mammoth mine on Bonita Creek in 1898. The Hart's Pass - Cady Mountain route was used during the early spring when the Azurite Pass route was impassable because of melting snow on its southern exposure. The 5 stamps from the Mammoth mine are now on exhibit at the Shafer Museum in Winthrop. (Horn-Olling Collection)

loosing its footing. A record-sized load was a graphite crucible, weighing 428 pounds. An even more troublesome cargo was a nine-foot camshaft taken over the Hart's - Cady Pass route, using an eight-foot sled. The 300-pound camshaft had to be reloaded at each of the switchbacks to keep the heavy side on the uphill runner.

The Kikendall brothers were known as tough men who knew how to survive in the wilderness. Don Olling knew them well and tells a story that illustrates their survival mentality. Don and his friend, Ed Kikendall, had been on one of the trips over Azurite Pass to the mine. They had traveled up the Methow River to the Horse Heaven cabin. This was a layover cabin built by the Kikendalls as a place to rest or stay during bad weather. It was located below the hogback which was the last leg of the trip to reach the summit of Azurite Pass. On this occasion, a chinook had blown in creating the danger of snowslides so they decided to hole up for a

Sumner (Stony) Stonebraker from Orofino, Idaho, showing a catch of fish. Two of his dogs watch with curiosity. Stonebraker housed about twenty-two of the animals in a shed at Robinson Creek above Mazama. Each had its own fenced enclosure. They were fed a menu of horsemeat mixed with bran and cod-liver oil. The high energy diet was healthy but caused the dogs to reek an offensive odor. (Photo courtesy of Jim Sparks)

time. The Kikendalls had always left a supply of grub in a fifteen-gallon galvanized barrel inside the cabin. Both men were hungry as bears and could hardly wait to dig into the store of supplies. But to their surprise, someone had gotten in and helped themselves to the grub and there was nothing to eat. Don asked, "Well, what are we going to do now?" Ed's reply was "Don't worry, we'll eat." They bedded down for the night and when Don woke up in the morning, he saw Ed unraveling a string of yarn from the top of one of his long wool socks. After he had pulled a good length of yarn, he began looking for something to use for bait. He found a few bits of bacon behind the stove on the floor. They went outside and as they started toward the Methow river (at this elevation a large stream), Ed asked Don to gather some rose hips to use as additional lure. When they got to the river, Ed tied the bait to the end of the yarn and dropped it into the water. The small trout were eager to grab on and when they did, their tiny sharp teeth hooked into the yarn just enough so he could slowly and carefully coax them to the edge. Ed told Don to stand on the bank and be ready to grab the fish while they were still in the water. In a short time, they caught about twenty fish which averaged around seven inches. After a good feed back at the cabin, they headed on up the hogback toward Azurite Pass.

During the winters of 1930-31-32, Dick Horn backpacked mail and supplies from Mazama to the mine. He snowshoed from Robinson Creek to Mazama, then back the same day. The next day he hiked from Robinson Creek over Azurite Pass to the mine, a distance estimated to be about twenty miles. The round trip was repeated about every two weeks. Once, Horn left the Robinson Creek cabin in the early morning and hiked all the way to the mine. Rather than stay over and return the next day, he embarked upon a difficut task - to hike out the same day. He crossed back over the summit and down the slopes of the Methow side, covering stretches of the trail he had hiked many hours earlier. To get down the steeper grades, Horn probably did as other snowshoers were accustomed to doing - allow gravity to take over. That is, slide on the snow twenty to thirty feet until the snow piled up in front, then climb over the pile of snow and slide another twenty to thirty feet. With daylight hours spent, the last leg of the return trip to Robinson Creek was in the dark. Beaten and dog-tired, Horn arrived back at the cabin at ten o'clock. He had achieved his goal - a round trip in the same day. And there was one additional task he was able to accomplish that day. He left the trail at a certain juncture and worked his way to a snowy ridge where he posted a mining claim. When it was over, he told a friend that his legs just about killed him. Few men today would attempt such a hike.

While making a mail delivery in January of 1931, Horn was caught in a snowslide which could have claimed his life. Using one of his broken skis as a digging tool, he was able to free himself from the snow. He sustained broken ribs and severe bruises, but was able to continue on to the mine. At the time, he thought he might not live to tell the story. Don Olling, co-worker and long-time friend, spoke of Dick as being one of the toughest men he ever knew.

The most difficult means of traveling to the Azurite mine was by "shank's mare" on the snow. Here, Dick Horn is shown with his sled and backpack. He is on webs (snowshoes). (Horn-Olling collection)

CHARLES BALLARD

SURVEYOR OF TOWNS AND DEVELOPER OF MINES

Charles Henry Ballard was the son of Dr. Levi Ward Ballard who graduated from Cleveland Medical College of Cleveland, Ohio, in 1844. In 1858, the doctor brought his two sons from his first marriage, Irving and William Rankin, and his third wife Mary (his first two wives had passed away) to the Oregon Territory where he practiced medicine for seven years. Charles was born on August 26, 1858, near Roseburg.

In 1865, the family moved north to the White River Valley. They settled where the town of Auburn is now located. Irving, William, and younger half brother, Charles, assisted their father in clearing the land to build a frame home. Dr. Ballard had considered moving to Seattle to practice medicine but there were already two doctors located there. A person of diverse abilities, the doctor served as postmaster of Slaughter precinct and ran a general store. He was a charter member of the White River Presbyterian Church and served as an elder. His view of the importance of the church in his community was evidenced by the donation of parcels of ground to both the Presbyterian and Methodist congregations. Dr. Ballard passed away on January 28, 1897.

Charles had an adventurous childhood growing up in the White River Valley. After his death in 1934, a niece, Mary Ballard McKee related some of the special times that he had as a child:

> *Charles' boyhood was spent along the banks of the White River, roaming the nearby forests and playing at war and hunting with the few white and many Indian children of the tiny settlement. He had many tales to tell of how they chased, and were chased, by the bears that were plentiful then. As a child he was ever curious as to what lay beyond the horizon, so when he was about sixteen years old he crossed the mountains with some of his dauntless companions, and crossed on horseback what is now Snoqualmie Pass and Blewett Pass. Though he was jeered at for carrying pen and ink on such a difficult trip, he recorded the journey in a diary which reads now, with its record of skirmishes with hostile Indians and of lost trails through uncharted mountain country, like a western thriller.*[31]

Charles Ballard, M.E., President and Superintendent of the Azurite Gold Company. (Azurite Prospectus)

In his memoirs written after the turn of the century, Charles told of numerous experiences he had while growing up. One humorous incident took place when he was quite young:

> *Father was a doctor and as they all did in those days, made his own pills; so I'd make bread pills and try to force the cat to take them. Once while the folks were away somewhere, I went into the henhouse, drove a setting hen off her nest, took father's best hat and poured in the powder, then proceeded to make pudding. Powder was very valuable those days, to say nothing of the eggs and Sunday hat, and when I was discovered at my new-style bakery, father took off a belt he wore, and gave me a sound thrashing, while mother begged him not to do it, as I was too young to be whipped so hard. I was about four then, but I think that must have been the beginning of my respect for my father's orders.* [32]

Charles' interest in mining also began at a very early age:

> *Father had taken up a ranch in the Grass Valley near Roseburg, Oregon, where I was born on August 26, 1858; they moved a couple of years later to Wilbur where we lived till I was 6. My father and brothers went to the mines in Idaho two different seasons, and when I begged to go along, he said I could when my hair got black. Since I was a towhead, I eagerly asked everyone how to get my hair to turn black; my interest in mining probably dates from then. Father often told us stories of the mines.* [33]

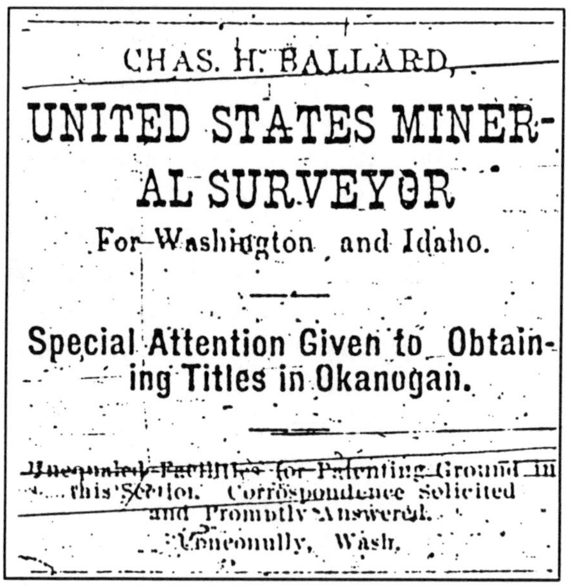

Advertisement in Okanogan Outlook*, July 22, 1892*

Charles Ballard was a brilliant student. While at home, his father tutored him in the art of diplomacy and the subject of mathematics. In his memoirs, he said that he was not required to study in the summer. However:

> *. . . Along in the fall, when the days began to get short and the nights long, I had to settle down to at least one regular hour and study arithmetic, and I had to learn whether I wanted to or not. None of the neighbor boys had to do this and for the first few winters I felt greatly aggrieved that I was not allowed to drop the book and go play when I wanted to. Firelight and homemade candles gave the light to study by. In the spring, lessons would be suspended and when fall came again I would be hoping that Father would forget or let it go. But some evening he would say, "Well, get your arithmetic and see how much you remember from last winter. "How much is nine times eight? "I don't know." "You don't know?" "Well, you knew the nines last winter. Now you sit right down there and see that you learn them before you go to bed." And I had to do it, for vine-maple sprouts and cedar shakes were handy and he had a way of using them sometimes to assist me when my memory got too bad.* [34]

As a young man, Charles learned the meaning of work. A lot of time was spent cutting firewood. Like the older men in the family, he had to use a man-sized axe which, to make matters worse, was usually dull. He got along quite well except the time when he was supposed to have cut enough wood on Saturday to last over Sunday night. It was the family policy not to work on Sunday. The Sunday that the wood ran out just happened to be the weekend that the preacher came to visit. Dr. Ballard had to go out on Sunday evening and cut enough wood to last overnight. This made him very unhappy. Years later, Charles still remembered his father's stern look as he told him he'd better cut enough next time, if he knew what was good for him.

Ballard's training, both formal and informal, provided the foundation for a future which would yield enormous accomplishments. In addition to his numerous managerial and scientific skills, he was respected for being a person of honesty and integrity. These qualities, too, must have taken root during his early childhood.

As part of his formal schooling, he attended the Territorial University in Seattle where he specialized in Civil Engineering. In addition to his personal studies in metallurgy, he took a special course in Chemistry and Assaying at Portland, Oregon. His first work as a civil engineer was in 1883 when he worked for the Northern Pacific Railroad making preliminary surveys through the western part of Washington. He also participated in the building of the Northern Pacific railway from Ellensburg to Seattle. Broadly speaking, this intrepid pioneer wore two hats, one as a surveyor and one as an assayer. As a surveyor, he platted several pioneer towns.

In 1886, Charles encouraged his father to plat part of the family homestead into town lots. The place was named Slaughter in honor of a Lieutenant Slaughter who was killed near there while on reconnaissance during the Indian war of 1855-56. In 1893, the name of the town was changed to Auburn. By that time, the town had grown substantially and some residents in the community were now wanting a name which they believed would be more acceptable.

Also in 1886, having recently received his credentials as a U.S. Mineral Surveyor, Charles, with his younger brother, Hazard, traveled to the silver mines in the Okanogan. Just prior to coming to the Okanogan, Charles worked as an engineer for the Anaconda Copper Company of Butte, Montana. The two brothers journeyed over a sagebrush trail which led through the place where the town of Wenatchee is now located. Their pack horses were loaded with food, personal items, and the numerous instruments which Charles intended to use as a surveyor and assayer. For the next several years he devoted considerable time to surveying mineral property and doing assay work. He once received a letter from the United States Mint Department complementing him on the accuracy of his work in assaying.

Ballard was present when the commissioners met to select a temporary county seat and temporary officials to serve for the newly formed Okanogan County. On March 6, 1888, the three designated commissioners and others met in a log cabin on the John Perkins ranch. After having taken official oaths before C.H. Ballard, a Notary Public, Guy Waring was elected chairman. C.B. Bash was chosen county auditor. Other officials and a county seat were selected the next day. Ruby became the county seat until a permanent seat of government could be

UNITED STATES
DEPARTMENT OF THE INTERIOR
GENERAL LAND OFFICE

PUBLIC SURVEY OFFICE
Olympia, Washington,
December 31, 1926.

Mr. Chas. H. Ballard,
U. S. Mineral Surveyor,
Twisp, Washington.

Dear Sir:

Your bond as U. S. Mineral Surveyor becomes renewable on January 7, 1927. Kindly advise me if you wish to renew your bond.

Very respectfully,
Geo. F. Naden
Office Cadastral Engineer.

U.S. Department of the Interior notice of bond renewal.

voted upon at the next general election. The voters then chose Conconully to become the county seat. Ballard was appointed Probate Judge by the Governor and held that post by election until the territory was admitted to statehood.

Charles laid out the townsites of Ruby, Conconully, and Chelan. The May 13, 1892, Okanogan Outlook announced that Charles had received notice of the passage and approval of the bill which was introduced in Congress confirming the government townsite of Chelan. The patent would be issued to Charles Ballard as trustee for the government. A six-month notice would be given for the filing of claims and at the end of that period a deed would be given for all lots except where there were adverse claims, which would have to be adjusted in the proper courts before deeds would be issued.

In addition to his other plats, Ballard drew an elaborate design for a townsite on the Columbia. The town of "Swansea" was to be located near the present-day town of Brewster. Plans for the town were described as follows:

> *Swansea was a great attraction on paper. Charles Ballard, a skillful draftsman and engineer produced the plat of the proposed city, which was recorded in 1892. Practically nothing was done toward building up the town of Swansea with the exception of the sale of a number of lots, regardless of the fact that the walls of nearly all the real estate offices in the country were decorated with blue prints descriptive of its attractiveness. Streets, alleys, and parks were profusely laid out, wharves of great capacity for rail and steamboat traffic covered the water front; the great Columbia was dotted with steamboats hastening to unload freight and passengers at the docks. To the brain that conceived and the hand that executed this work, too much praise cannot be accorded. Of course, all of this was conducive to the sale of quite a large amount of real estate. But owing to disagreement between two partners interested in the project, the enterprise of building up the town of Swansea was abandoned.* (35)

Charles and Hazard did a lot of traveling throughout the country to meet with mining personnel and investigate properties. In the spring of 1894, Charles made a trip down to the Columbia River placers to experiment with a new gold-saving machine which had been invented by S. Barnhart of Ruby. His opinions about methods and equipment were no doubt highly respected among those in the mining industry. "Have you checked with Charlie Ballard?" was probably a question frequently asked by miners.

The Ballards were frequently in the news around the Okanogan in the 1890's. In January of 1892, Charles traveled to Olympia to attend a meeting of the executive board of the Washington World's Fair Commission, of which he was a member. He had obtained a large supply of ore sacks which were to be distributed free to miners and prospectors who were willing to contribute to the World's Fair exhibit representing the Okanogan. At that time, Ballard and S. S. Collins owned a 34'x50' building on Maude Avenue in Conconully, which was rented for business purposes.

Charles Ballard was interviewed by the Spokane Review in January of 1892. He had been staying at the Hotel Spokane and was scheduled to travel to Olympia to meet with the executive board of the Washington World's Fair Commission. He gave an update of happenings around the Okanogan:

> *You heard, I suppose, said he, of Mr. Dunn and a company of Seattle capitalists bonding the Lakeview and Mammoth mines on Mineral Hill, just west of Conconully, for $30,000. They have put a force of men to work on them already under the charge of Mr. Mathewson, and I understand the force will shortly be increased considerably. The same parties have recently acquired large interests in the Chelan district and in the Monte Cristo and other camps on the west slope of the Cascades. They are said to represent a great deal of capital, and their advent to the country cannot fail to prove of enormous advantage to it.*

Mr. James F. Wardner has put a force of men to work on the Red Jacket mine in Toats Coulee, about 17 miles from Conconully, which he recently bonded for $20,000. This mine was only discovered last fall and is a rich gold property, the mines of Ruby and Conconully being silver propositions and principally chloride ores. It is only since the discovery and development of the Wannacut Lake district that people have began to consider Okanogan county a likely gold producer.

The air is full of rumors regarding mills to be put in the spring. I am informed on good authority that a large custom-mill will be erected in the neighborhood of Ruby or Conconully for the treatment of silver ores ... On Palmer mountain there is a group of mines which in spite of all these drawbacks is going right ahead. The Ivanhoe group is composed of silver productions only, and they have made several good sized shipments. The last, which was to the smelter at Tacoma, consisted of about 10 tons of ore and netted the owners something over $3,000. What a saving could have been effected had there been a mill in the camp to treat this ore instead of paying freights on it by wagon and railroad all the way to Tacoma ... There is one thing, however, about which we are all doing some kicking, and that is the fact that we have no direct stage line or mail route from our section to the land office at Waterville. To get from Conconully or Ruby to Waterville we have to go around by Coulee City and back again to Waterville, or else hire a conveyance at Port Columbia and drive from there. A petition has been got up and forwarded, asking to have this route established, and if it is done it will then be possible to take in Waterville either in coming out or going in to the country without wasting three days or more.

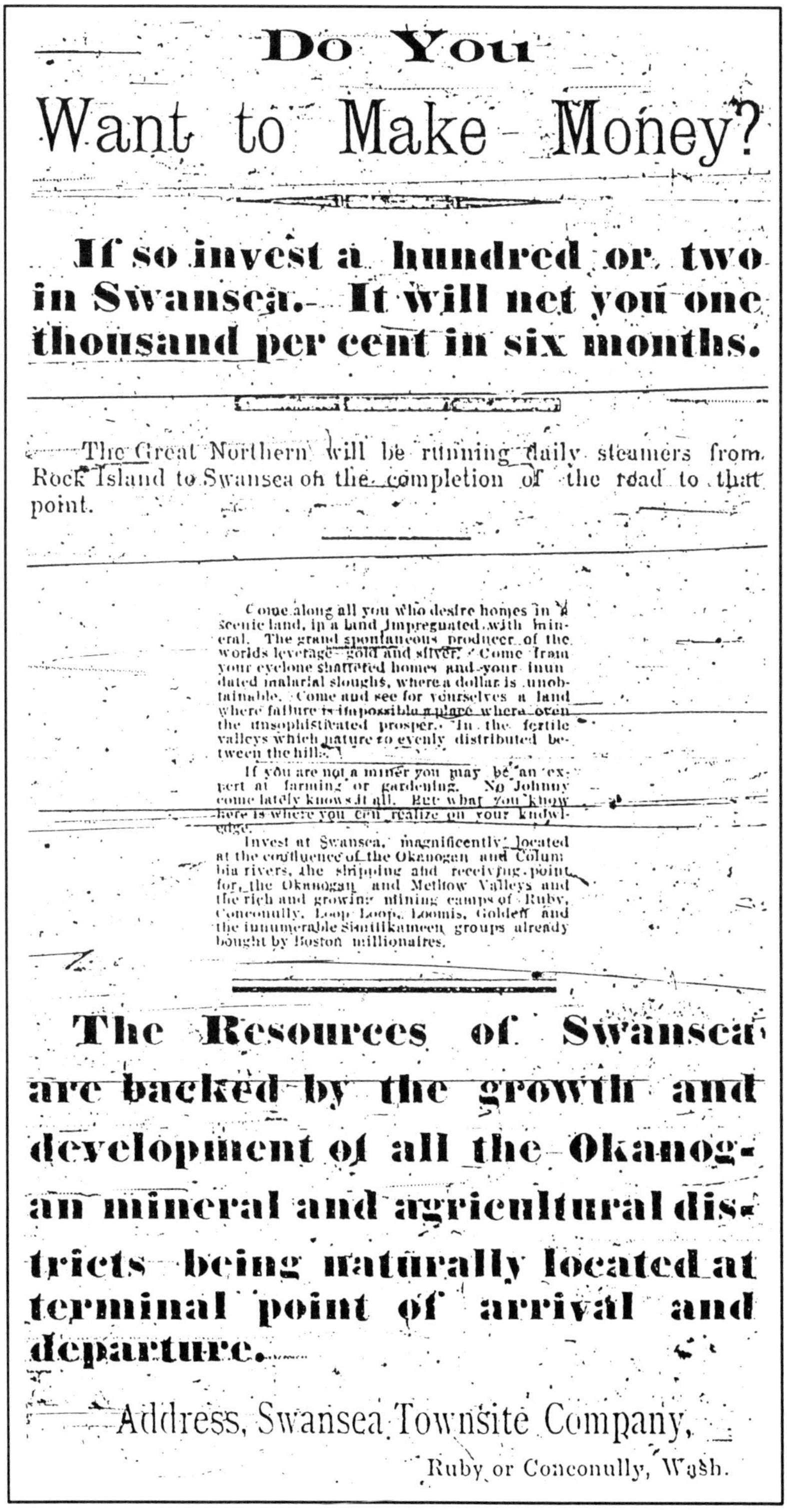

Do You

Want to Make Money?

If so invest a hundred or two in Swansea. It will net you one thousand per cent in six months.

The Great Northern will be running daily steamers from Rock Island to Swansea on the completion of the road to that point.

Come along all you who desire homes in a scenic land, in a land impregnated with mineral. The grand spontaneous producer of the worlds leverage—gold and silver. Come from your cyclone shattered homes and your inundated malarial sloughs, where a dollar is unobtainable. Come and see for yourselves a land where failure is impossible a place where even the unsophisticated prosper. In the fertile valleys which nature so evenly distributed between the hills.

If you are not a miner you may be an expert at farming or gardening. No Johnny come lately knows it all. But what you know here is where you can realize on your knowledge.

Invest at Swansea, magnificently located at the confluence of the Okanogan and Columbia rivers, the shipping and receiving point for the Okanogan and Methow Valleys and the rich and growing mining camps of Ruby, Conconully, Loop Loop, Loomis, Golden and the innumerable Similkameen groups already bought by Boston millionaires.

The Resources of Swansea are backed by the growth and development of all the Okanogan mineral and agricultural districts being naturally located at terminal point of arrival and departure.

Address, Swansea Townsite Company,

Ruby or Conconully, Wash.

Advertisement in *Okanogan Outlook, July 22, 1892*

THE BALLARD BROTHERS

As communities were still being carved out of the wilderness, other children of Dr. Levi Ballard were earning reputations as prominent businessmen and leaders. As already mentioned, Hazard assisted Charles in the early development of the Azurite mine. For a number of years, Hazard operated a sawmill with his father-in-law, Hank Johnson. The mill was located at a place called Rockview on the Methow River, about half way between Mazama and Winthrop. Hazard and Walter Paselk discovered the Gold Hill mine which was located on the opposite side of the mountain from the Azurite. The original claims at the Gold Hill were recorded by Hazard Ballard et al. on November 6, 1930. Hazard organized the Gold Hill Mining Company and with the help of Paselk began development work. Now the two Ballard brothers were engaged in working a vein system that was thought to have continued all the way through the mountain, from the Azurite to the Gold Hill.

The silver-gold ore samples from the Gold Hill mine were promising. Paselk, who worked as Hazard's foreman, spoke of the "startling veins." He concluded, "This is the most wonderful gold prop-

The Ballard brothers, left to right - Arthur, Hazard, Leon, and Charles, the eldest of the four brothers. (Courtesy of Okanogan County Historical Society)

erty I ever saw in all my prospecting and mining." Thousands of shares of mining stock were sold to hopeful investors. One optimistic purchaser bought 22,237 shares. By the summer of 1938, over 3,500 feet of tunnels and crosscuts had been driven into the mountain. To get the ore out of the hills, a "mine to market" road was planned down the west slope to Diablo Lake. The ore was to be barged down the lake and then taken by rail car to the Tacoma Smelter. Five miles of the road was completed in 1938. Promoters hoped to receive matching funds from Whatcom County which would enable them to finish the remaining portion the following year. When funds did not come through, the road project was postponed and eventually abandoned. [36] The problem of transportation of ore was one of the unresolved issues which ultimately resulted in the closure of the mine.

Hazard and his wife, Zora, lived on a small farm at Lost River about seven miles above Mazama, Washington. Hazard was a large man, over six feet tall, and was known for being a horseman. Over the years he owned and cared for a sizable number of pack horses and a few mules. The animals were used mostly for packing into the mines, although at times he packed for the forest service, hunters, and fishermen. At one time, in 1933, he owned about fifty head which were stabled at Lost River. At Thanksgiving time they were taken to the Okanogan for wintering, then brought back in April to be shod

NUMBER

SHARES

Gold Hill Mining Company

WINTHROP, WASHINGTON

NON-ASSESSABLE

This Certifies that S. N. Shafer is the owner of One thousand Shares of the Capital Stock of

Gold Hill Mining Company

transferable only on the books of this Corporation in person or by Attorney upon surrender of this Certificate properly endorsed.

IN WITNESS WHEREOF, the said Corporation has caused this Certificate to be signed by its duly authorized officers and its Corporate Seal to be hereunto affixed this 22nd day of Oct A.D. 1935

R. M. Badger President

Secretary

SHARES ONE CENT EACH

A Gold Hill mine certificate. Simon Shafer, a Winthrop businessman, purchased 1000 shares of Gold Hill stock. The Shafer Museum, located at the Guy Waring log home above Winthrop, was started by Simon Shafer. Shafer purchased the home in 1943. Over the years he collected artifacts from around the country to add to the display. In 1976, Simon's children presented the museum to the Okanogan County Historical Society. The museum is a major attraction for the thousands of tourists who visit Winthrop each year.

and readied for the summer season. Each year around the first of May, Hazard took the horses over Hart's Pass to deliver fresh supplies to the mines, sometimes through deep snow. When the horses got bogged down and couldn't move, the men had to shovel them out.

Bill Wehmeyer, Harley's son, helped his dad with the horses during the time they were packing in and out of the Gold Hill mine. In the fall of 1935, a winter's supply of groceries, 5,000 gallons of gasoline in five-gallon cans, fifteen tons of dynamite, ore cars, and other machinery parts were packed to the mine from Riprap, a distance of about eighteen miles. Another packing outfit delivered some of the supplies and machinery from Riprap to a barn located at the junction of Granite and Canyon Creeks. Hazard's horses carried the cargo the remaining distance to the mine.

Handpicked pieces of ore from the Gold Hill were taken by packtrain to the barge landing at Riprap. Each horse carried two sacks of ore and each sack weighed 125 pounds. When loaded, the horses didn't appear to be carrying much. The sacks were small (about the size of sand bags) but the silver-gold ore was very heavy. Ore was also hauled down to where Ruby Creek empties into the Skagit River, then up the Skagit to be unloaded at a place called Boundary, near the border. From there it was taken into Canada, possibly Hope, to be smelted. The third route used for transporting ore from the mine was over Mebee Pass, then down the Methow to Pateros. In 1936, 3,680 pounds of hand-selected ore was taken to Kellogg, Idaho. The net value of the shipment came to $84.42.

A blacksmith at Marblemount built a horsedrawn vehicle to transport to the mine the blocks for the compressor (3-horn rig) and motor, each block weighing about 400 pounds. He put together odds and ends of whatever was available. A Model T Ford wheel was placed onto an axel. The axel was affixed to shafts which could then be harnessed to a horse. A set of steel plow handles were attached at the rear for stabilizing the load. By much maneuvering and strenuous effort, the blocks were delivered

Hazard Ballard's pack string on Boulder Pass at the head of Boulder Creek. The horses and two mules, about twenty total, had been brought from Lost River over Hart's Pass, down Slate Creek and Canyon Creek, then up Boulder Creek. From Boulder Pass, they made the descent to the Gold Hill mine located above East Creek, a branch of Granite Creek. The trip took two days and covered about thirty miles. Sometimes, twenty-seven animals were run in a string and were trained to lead without being tied together. Harley Wehmeyer is on the horse in the middle. Hazard is at the rear. The forward horses were Dixie and Tom. They were taking supplies to the Gold Hill mine where cabins were being built at the time. (Photo courtesy of Bill Wehmeyer)

to the mine. Much of the trail up East Creek was steep with numerous switchbacks, a challenge for man and animal even without the extra weight of cargo, especially a 400 pound load perched on a single wheel. The recollection is that the wheel could not be steered. One can imagine how difficult it would have been getting around the sharp corners. It took four days to deliver each of the blocks. Later, during the winter, a young man who delivered mail to the mine carried a seventy-five-pound anvil all the way from Riprap to the mine, quite an undertaking.

Another brother of Charles, Leon F. Ballard, served as Notary Public at Conconully in 1893. The April 2, 1892 Okanogan Outlook printed public notice of several mine patents being applied for by Leon Ballard whose mailing address was Slaughter, Washington. Among the twelve mines listed with survey descriptions were the Iron Cap, Welcome Nugget, Gold Cup, Silver Spray, Silver Strike, Sullivan, and Silver Flood. In 1911, Leon accompanied Charles to Alaska where Charles worked at assaying gold bullion for banks and the United States Assay Office. For a number of years, Leon lived with Hazard and Zora at Lost River and assisted them with work on the farm. They were known for the boysenberries they raised and sold. For a time, Leon took care of the base camp at Robinson Creek where he cooked and helped with the freight.

The youngest of the four brothers, Arthur C. Ballard, served as city clerk for the town of Auburn in the 1920's. He had attended Whitworth College and graduated in 1899 from the University of Washington where he majored in Latin. He became renowned as a lay anthropologist and his work with the local Native American language and culture earned him the respect of professionals in the field. Among his writings were "The Mythology of Puget Sound," "Some Tales of the Southern Puget Sound," and "Calendric Terms of the Southern Puget Sound Salish." Arthur served as secretary of the Azurite Gold Company. Charles was the president and general superintendent. He must have shared Charles' interest in mining as early as 1899. A personal letter discovered in Tacoma and dated January 25, 1899, made comment regarding the fact that Mrs. Levi Ballard had rented out most of her home to a family and that Charles and Anna had been there to visit for a week. When they returned east of the mountains, Arthur went with them. At that time, Charles operated the Mammoth mine in the Slate Creek District. Arthur worked at the Mammoth at least one season.

A half brother, Irving, had been an attorney of King County prior to his death in 1880. Another half brother, William Rankin, had the honor of having a town named after him - Ballard. William attended the academy at Wilbur, Oregon, and Washington State University. In 1876 he accepted the position of mate on the steamboat Zephyr which plied between Olympia and Seattle. At that time, the boat was owned by a brother, assumed to be Irving. In 1877, William was made captain of the Zephyr so then became known as "Captain Ballard." In 1881, he became a part owner with two other persons and in 1886, the sole owner. He sold the boat in 1887 to the Tacoma Mill Company. [37]

In May of 1882, William, in partnership with Judge Thomas Burke and John Leary, purchased 720 acres of land which were platted into ten-acre tracts. It was named the Farmdale Homestead. In 1887 the tracts were marked off into town lots and water-front mill sites. It then became known as Ballard Junction, named as such by the manager of the Seattle Lake Shore and Eastern Railroad which serviced a line through the village. In 1890 the town was incorporated, bearing the name of its founder - "Ballard." The town ceased to exist as a municipality when annexed by the City of Seattle in May of 1907. The community would still be known as "Ballard." William R. Ballard became known as the "Father of Ballard." [38]

William, like Charles, had been trained as a civil engineer and worked as a surveyor. In addition to subdividing and selling land on Salmon Bay, William had been awarded government contracts for surveying public lands, including the Yakima Indian Reservation which took three years. In the 1890's, he became a prominent figure in the banking industry of the Seattle area. He was also president of the First National Bank of Waterville, Washington.

THE KNOB HILL MINE

A Success Story

The historic Knob Hill mine, owned and managed by the Hecla Mining Company, is located near the town of Republic in northeast Washington. The history of the Republic Mining District dates back to 1896, the year that the North Half of the Colville Reservation was opened to mineral entry.

The Knob Hill mine went into full production in 1902 and has operated almost continuously since that time. The Knob Hill No. 2 incline shaft, first sunk in the late 1930's to reach deeper veins, is 2,083 feet in length and is used for hoisting all ore and rock from the Knob Hill and Golden Promise deposits. The hoist capacity is 13,000 pounds and the speed is 815 feet per minute.

In 1984, the Golden Promise system of ore was discovered. This has proven to be one of the highest grade ore deposits to be found in the Republic Mining District. The new Golden Promise shaft, sunk in 1986, is 1,300 feet deep, seven feet in diameter with steel lining and provides the main access for workers and supplies. The Golden Promise connects with the historic Knob Hill by a haulage tunnel which is more than one mile long. Since 1941, when the first recorded ore was mined and hoisted up the No. 2 shaft, more than two million ounces of gold have been recovered from the mine.

Built in 1937, the existing mill at Hecla's Republic Unit uses flotation and cyanidation methods to recover precious metals from the ore. The mill processes approximately 325 tons of high grade ore per day, yielding a recovery rate of 90% of the gold and 80% of the silver. [39]

The mill at Hecla's Republic unit, formerly the Knob Hill, near Republic, Washington.

Working underground at Hecla's Republic Unit near Republic, Washington. (Courtesy of Hecla Mining Company and VanGundy Photography, Spokane)

Author's Note:

As of the summer of 1998, the Knob Hill mine was shut down. Workers were in the process of closing all underground openings, re-claiming the tailing impoundment, and removing the mill building and other surface structures. There was no projection of future underground workings.

Republic, Washington in 1914. The population at that time was 1,500. (Plate XI from U.S.G.S. No. 550, 1914)

The Republic mill, 1914. The Republic Mining District, formerly known as the Eureka District, is located in Ferry County about twenty-five miles south of the International Border. The district was actively prospected after the opening of the North Half of the Colville Reservation on February 21, 1896. On March 5, Philip Creasor and Thomas Ryan staked out the Republic claim southwest of the present-day town of Republic. The total estimated production for the district up to January 1, 1912, was $5,000,000; 83% represented gold and the remainder silver. A chief producer was the Republic mine. Other important producers were the Knob Hill, Insurgent, Imperator, Quilp, Sanpoil, and Hope. (Information and Plate XVI from U.S.G.S. No. 550, 1914).

PART III
PLACER MINING

Placers are deposits of sand, gravel, and other alluvium which contain valuable minerals that, in most cases, may be worked by washing or amalgamation. Placer gold, then, is gold which has been released from its host bedrock by weathering or abrasion. A placer deposit is formed when valuable minerals are concentrated either in place (near the bedrock source) or at some distant location.

Various types of placer deposits have been identified by their location. A residual placer is one which is located in the immediate vicinity of the original lode source. The released minerals are concentrated in place. Some prospectors have referred to residual placers as "seam diggings" since gold-producing veins have been found directly beneath the deposit. This was the case of the "Glory Hole" ledge found by Alex Barron in 1893. Transported placers such as hillside, bench, and stream, are found a distance removed from the original bedrock source. For most prospectors, river and stream placers are considered to be the most productive. The ongoing process of erosion continues to enrich some of these placers.

Early prospectors were always looking for signs which would lead them to the vein from which placer gold originated. If they found traces of gold that increased in size and roughness as they advanced up a stream or hillside, it often meant they were getting close to its source. Also, experienced prospectors were interested in "float" - that is chunks of ore eroded from a gold vein. Sharp and angular pieces usually indicated they had not traveled a great distance. By following the trail of float, gold-producing veins were sometimes discovered.

Placer gold has been classified into several sizes. Those in prospecting circles have referred to gold in four general categories: flour, fines, coarse, and nuggets. Flour is just that - gold so fine that it resembles flour. It can take thousands of particles to make a grain. When dry it will float on water. Fines would refer to larger flakes and pieces which can more easily be gathered into measurable amounts. The methods by which a prospector recovers fines or flour gold is a science in itself and the tools used in this endeavor range from the simple to the complex. Coarse gold consists of pieces about the size of a grain of wheat or larger. Sometimes, coarse gold is considered as being small nuggets. It's been suggested that anything larger than one grain in weight is a nugget. Nuggets have been categorically sized into small ones and large ones, clinkers and clunkers. A clinker would be one that goes "clink" when dropped into a gold pan and a clunker would be one that goes "clunk". Then there are those unusually large chunks which are in a category by themselves, such as the Welcome Stranger found in Australia in 1869. It weighed a phenomenal 2,280 troy ounces.

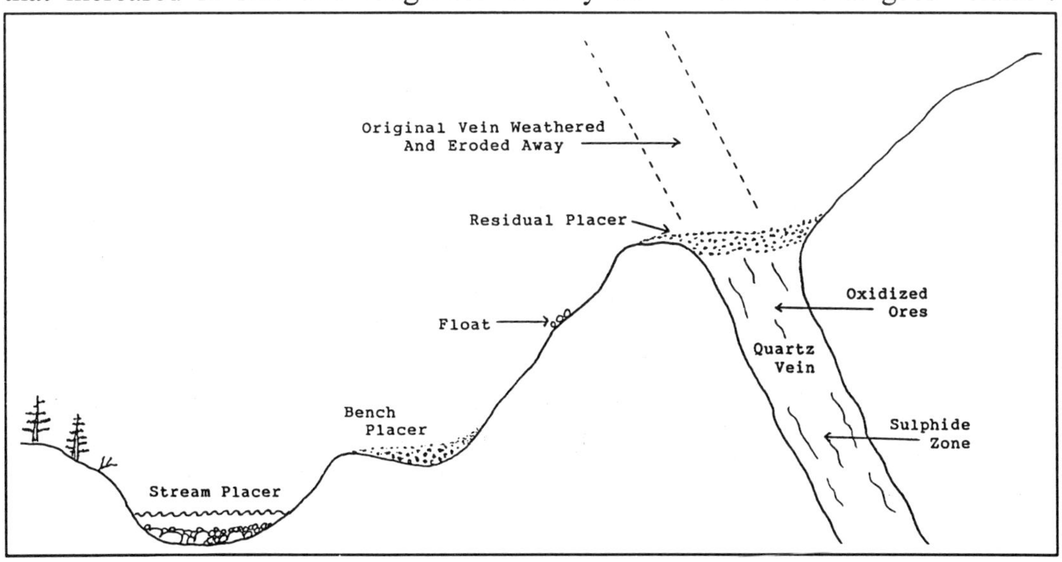

THE PROPERTIES OF GOLD

Gold is extremely malleable and ductile, a factor of importance for both dentists and jewelers. On the Moh scale, from 1 to 10, it has a hardness of 2.5. In comparison to gold, iron pyrite has a hardness of about 6.25. Pyrite, which is sometimes confused with gold because of its pale yellow color is brittle and most often has angular edges. Placer gold usually has worn, smooth edges, depending on the distance it has traveled. The differences between gold and pyrite are quite apparent once a person has compared the two.

Gold resists corrosion and is largely unaffected by the ordinary acids of nature. For this reason it is considered a "noble" metal. Even after centuries of exposure to the harshest of nature's elements, it emerges with distinguishable beauty. At a mere glance, the experienced prospector can recognize even the smallest particle of raw gold and affirm - "Yep, that's gold!"

In addition to gold's enduring quality, another feature of importance to the early and present-day prospector is its specific gravity. Specific gravity is the weight of a substance compared with that of an equal volume of water. Pure gold possesses a specific gravity of 19.3. As found in nature, it weighs

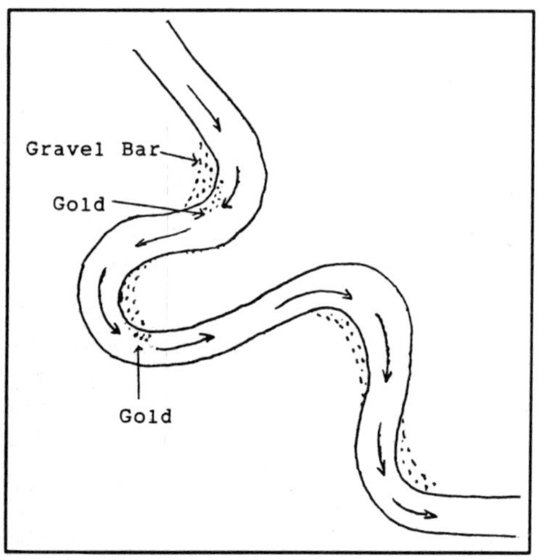

During heavy flooding, gold is moved along with other rock and debris. It will follow the main path of the current and have a tendency to concentrate in places where the current slackens, such as below the inside curve of the river. I have found some of these areas to be very productive.

a little less because of lighter metals alloyed with it. Gold is about 2½ times the weight of iron and eight times that of common sand. Because of its weight, when agitated or moved about in surrounding material, gold particles will usually migrate downward. Historically and today, most systems used for the recovery of placer gold take advantage of the comparable weight of gold to other ordinary sands and gravel. This would include any of the traditional tools such as the gold pan, rocker, and sluice.

WHERE THE GOLD IS FOUND:

Because of the weight of gold, there are certain patterns of travel to be considered when prospecting. Books have been written on the subject of where to find placer gold. One rule frequently suggested is that gold settles wherever the stream current slows, such as on the inside curve of the river or below a narrow stretch with fast current. As the current slows, the gold drops to the bottom and concentrates in the immediate area. During one summer when I was dredging with friends on the South Fork of the Forty Mile in Alaska, a very rich paystreak was found on the inside curve of the river. The gold probably originated from a heavily eroded bench which ran along the river for a considerable distance upstream. The swift current carried the gold over a stretch of bedrock depositing it onto a layer of hardpan rock and gravel - a natural gravity trap. The gold fines lay within a few inches of the surface. During that summer many ounces of gold were recovered from this short stretch of the river.

Another place frequently suggested is in an ancient stream channel, sometimes located on the hillside high above the water line. One may find streaks or pockets of black sand between the rocks of the original streambed or on top of a layer of clay. Black sand is always a good indication of gold being present. As a rule, gold is usually found in accompaniment with black sand which is composed largely of magnetite (magnetic iron oxide). A small magnet will pick up magnetite. So, if black sand is not found, there probably will be no gold unless an isolated nugget with quartz matrix happened to have been washed along with other rock. Such prized specimens have been found with a metal detector.

Gold has been found concentrated at the base of large boulders lying on or near bedrock. During flood waters when gold is moved along with other debris, the gold has a tendency to settle out around any obstruction that slows or alters the direction of the current.

In prospecting for placer gold, a person should always keep in mind the weight of gold and the principle that it will usually seek the lowest possible location. Generally speaking, it will rest where something impervious stops it, such as clay, hardpan, or bedrock. Look for those natural gravity traps, such as small crevices in bedrock, where the gold can be caught. Once a nugget drops into a crevice, it will remain there until some prospector or act of nature dislodges it from its bed. There are prospectors who spend nearly all of their time in the field searching bedrock seams for the heavier and more valuable chunks of gold.

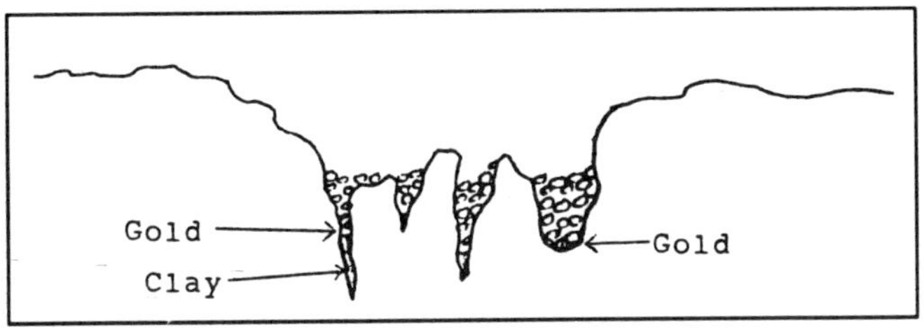

MORE LIKELY!
A hole with crevices or broken bedrock in the bottom is more likely to catch and hold gold than one with a smooth surface. Some of the nicest nuggets I have found were embedded in the cracks of bedrock, usually laying on clay and beneath compacted rocks. As I recall, I have never found gold in or under clay. Any crevice or crack in bedrock, either at the shore or on the riverbottom, is a good place to look for gold. A teaspoon or screwdriver are ideal tools for digging between the rock.

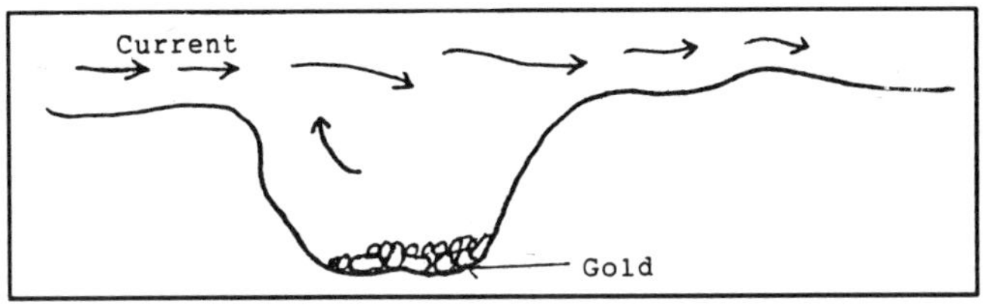

LESS LIKELY!
A "boil hole" with a smooth bottom and sides is considered an unlikely place to find gold. The theory is that nuggets which remain in this type of hole will in time be ground to flour by the churning action of rocks. The fine particles are washed downstream. However, one should never discount a pothole, especially if it has a layer of compacted rock and gravel in the bottom. It should be checked out. Gold may be found beneath the rocks on hardpan, bedrock, or clay. While dredging, I have found a few nice nuggets under compacted rock at the bottom of a hole. On a few occasions, too, I have found what appeared to be recently deposited gold (small amounts) on smooth bedrock with loose rock and gravel. The bottom line is - gold is where you find it!

A few days of working bedrock on the Similkameen River has yielded a nice assortment of gold. Because the river has been heavily prospected for nearly 140 years, gold as shown is not easily found. Similkameen gold is distinctively beautiful, orange in color.

THE SIMILKAMEEN RIVER

The headwaters of the Similkameen River gather in the high elevations of the eastern slopes of the Cascade Mountains, from both above and below the Canadian border. Flowing south and crossing the 49th Parallel under the shadow of majestic Mount Chopaka, the river continues its journey around the "big bend" from which it turns east toward the Okanogan Valley. As it approaches the valley, dropping fast, the river becomes a millrace of rapids. The downhill torrent then becomes quiet as it merges with the Okanogan River just below the small town of Oroville. From here the waters of the Okanogan meander down the valley, providing the natural course of the old Cariboo Trail, to finally merge with the Columbia River near where the historic Hudson's Bay Fort (Fort Okanogan) was once located.

The modern history of the Similkameen goes back to 1859 when men of the Boundary Commission surveying the International Line discovered gold along the river:

> *They were bound south and totally ignorant of the character of the territory they must needs traverse, determined beside the beautiful and clear Similkameen to camp and recuperate preparatory to the commencement of their long journey to the southward . . . for they had been in camp scarce a fortnight when they were electrified by the discovery of golden metal in abundance, made by their number, who, noting the sparkle in the gravel lining the banks of the stream, became speedily convinced by investigations he made that gold lay along its shores in abundance.* (1)

The November 18, 1859, Pioneer and Democrat published a letter which had been received from Fort Colville. The letter was dated October 21, 1859. It began by saying that several buildings were being constructed to provide winter quarters for the Boundary Commissioners who had been on the Similkameen. The letter continued:

> *Captain Archer's command are expected to return about the first of November. Reports well authenticated and generally believed here, say that the command have discovered good paying gold mines on the Similkameen River, and are busily engaged in mining. One of the party found a piece weighing $27.00, others have found pieces weighing from $5 to $15. Thomas Smith, formerly of Vancouver, and government expressman between these two points, who comes and goes once a week, reports the mines to be good. Mr. Holley arrived from the mines a few days since bringing with him several pieces which he exhibited. Those which I saw resemble grains of wheat. He reports that as yet, only paying dirt is found upon a bar in this river. The miners are sanguine that further prospecting will develop rich mining.*

The bar referred to in the letter may have been the famous "Rich Bar."

Another letter from the Dalles, dated November 7, 1859, related information received from members of the Boundary Commission who had recently arrived there. Employees had brought with them samples of gold from the river. They reported that prospects had been made five miles each direction from the bar on which gold was first discovered. Miners were recovering 50¢ to the pan, that is the "frying-pan." Mining implements being scarce, prospectors resorted to using common cooking pans. Even by these means, they were getting a good showing of color.

In January, 1860, a miner wrote that seven miles along both sides of the river had been determined to be rich with gold and that cold weather had prevented them from prospecting any further. The thermometer at Fort Colville had dipped to twenty-three degrees below zero. In spite of the extremely cold temperature, seventy miners had been wintering on the river. It was so cold that "if they attempted to wash a pan of dirt, the pan would freeze to their fingers." Nothing more would be known about the diggings until the snow disappeared in the spring.

As soon as word of the discoveries reached the outside, hundreds of prospectors flocked to the Similkameen. In 1860, a gold mining town of tent dwellings known as Okanogan City sprang up along the river. The settlement was short-lived, as miners soon moved to more promising locations in Canada. After a few months, though, many returned to the rich values of the Similkameen.

RICH BAR ON THE SIMILKAMEEN

A huge amount of gold was mined from the Rich Bar stretch of the river. It's still a matter of speculation as to where the famous bar was precisely located. If Rich Bar was also the discovery location as mentioned in a letter of January, 1860, and described as being "seven miles from the crossing of the Similkameen," then it must have been somewhere in the general vicinity of Eagle Rock. Also, since a wing dam is believed to have been built at the location giving miners access to the gold, then bedrock must have been close to the surface. Most of the heavy gold found on the Similkameen has been on bedrock, which it should be added, has been well worked over the decades.

Eagle Rock, a landmark on the Similkameen, is about a mile above Shankers Bend. Some believe the famous Rich Bar was somewhere in this vicinity. In the 1930's, miners cordoned off the river at this location to work the riverbottom but were surprised to find very little gold. (1930's photo - courtesy of Art Gjerde)

Though it has never been located, prospectors have theorized that a mother lode was the source of

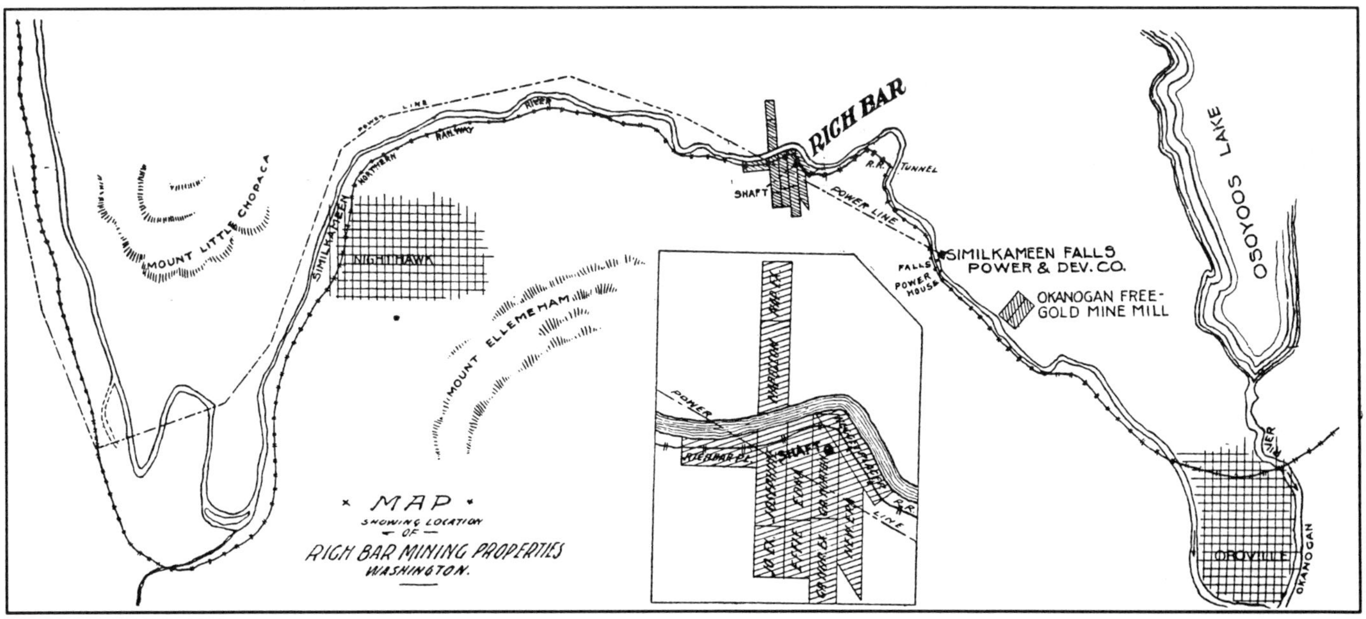

Map from a 1909 promotional brochure advertising the Rich Bar Mining Company. The hardrock and placer claims as shown were located on the river just above the Shankers Bend. The claims were in the general vicinity of the (placer) Rich Bar which was discovered around 1860.

the Rich Bar deposit. Supposedly, by the perennial process of erosion, the coarse gold was released from its lode vein and washed downward to the river. In spite of hardships of immense proportions and tools of simple nature, those lured to this remote place did collectively harvest an incredible amount of gold. One can imagine the excitement as the first prospectors rumored their fabulous finds. Various estimates have been given as to the amount of gold taken from the river during the gold rush years. One early report (frequently quoted) suggested that approximately $500,000 in gold was mined from the Similkameen placers in the immediate years after 1859. During the peak of activity 3000 miners are said to have worked the stretch between the mouth of the river and Similkameen Falls where Enloe Dam is now located. Decades later, the January 6, 1892, Spokane Review reported that mining was still being carried on with satisfactory results along the river. Using sluices, miners were making from $3.00 to $3.50 per day, "and occasionally a rich pocket is struck where much better pay is found. . . ."

EARLY ROUTES OF TRAVEL

The Dalles

The February 10, 1860, issue of the Pioneer and Democrat presented recommendations from several sources as to the best routes to be taken to reach the Similkameen and upper Fraser River. The longest and southern-most route was from Portland through the Dalles. A miner had investigated the route during the "excitement" of 1858. With hundreds of others, he drifted by the "only true natural channel," to the city of Portland. What he encountered was less than congenial. In his opinion, the miners were being taken advantage of by those providing services and supplies:

> *. . .We were besieged by a numerous delegation of those "fine citizens," who, in the warmth of their disinterested friendship, grew eloquent in the praise of their particular route, via the Dalles; but invariably would up their philippics with sly recommendation to purchase our supplies of some particular house in that place, or some of the upriver towns, as each was individually interested. No assertions were too bold, nor means too foul to be employed to carry their point, and that, too, in the face and eyes of hundreds of honest men, who had been "just clean through the mill" and had come back completely fleeced of their hard earnings, to hunt their way, as they well might, among a more hospitable people and over a more feasible route to their destination.*

Snoqualmie Pass

In the paper's discussion of the travel routes, the most favored route to the Similkameen was from Seattle over Snoqualmie Pass. Seattle was a small town on the Sound with access for the largest steamers and sailing vessels. The trail over Snoqualmie was described as having been improved with easy grades and passable log bridges. It was practicable for mules carrying from 300 to 400 pounds each. Everything needed for a miner's outfit could be purchased at reasonable cost from several stores located in the Seattle area. After crossing the Snoqualmie summit, prospectors were advised to follow a specific course of travel:

> *. . . to upper Yakima valley; take a turn to the north across a high mountain until you reach the Wenatchee River; cross by canoes, and then follow the trail northward to the Methorn (Methow) River; then follow the*

bank of the Columbia to Okenagan; thence up the Okenagan to the Similkameen. Seven miles from the crossing of the Similkameen is where the first gold was discovered. . . .

Several things may be noted regarding the 1850's use and improvement of the Snoqualmie trail. Indians were known to have crossed and re-crossed the pass during all seasons of the year, even in the winter. During mid-winter, a Northern Pacific Railroad Survey party had crossed the pass from the eastern to the western base of the Cascades. In the summer of 1858, seventeen citizens from King County spent several weeks improving the trail, removing obstacles and shortening its direction. After the improvements, Mr. Yesler and Mr. Denny and several others from Seattle traveled over the trail to determine its condition. They reported that they had no trouble and that it was in excellent shape. In August of 1859, community leaders of King County held a meeting to discuss the building of a wagon road over the pass. Captain A.C. Rand chaired the meeting, and Messrs. H.L. Yesler, D.S. Maynard, Franklin Matthias, and A.A. Denny were elected to collect funds and generally superintend construction of the proposed road. A subscription was circulated and $1,050 was immediately raised. As of February of the following year, citizens of Seattle and neighboring communities had spent several thousand dollars for the building of a good wagon road over Snoqualmie which was expected to be completed in the spring.

The Whatcom Trail

In the February 10, 1860 discussion of the routes, miners were advised not to attempt traveling the supposed Whatcom Trail. Citizens in that region had unsuccessfully spent a large sum of money to discover an opening through the mountains. A negative appraisal was candidly expressed:

Concerning that trail, we are frank to confess we were "humbugged" about as much as the too incredulous but confiding miner, who patiently waited from week to week to hear of the trail's completion, and consumed his grub on Whatcom's shores.

Two years earlier, during the spring of 1858, the proposed trail which was to travel across the Cascade Mountains to the mouth of the Thompson River in British Columbia was a subject of intense public interest. A fund for the construction of the trail beginning at Whatcom had been raised by subscription and the task of heading the project was assigned to Captain W. W. DeLacy. Progress was being made but the trail was not expected to be completed for several weeks. Merchants knew that when it was completed, they would prosper from the trade generated by a steady stream of prospectors. Zealous promoters praised the benefits of the trail as though it were already open for travel, thus encouraging some to prematurely head out.

At one time, in April, 1858, there were an estimated 1,000 persons crowded together in the towns of Whatcom and Port Townsend. During the following weeks, miners continued to arrive from Oregon and California. By the end of May, Whatcom had grown from a few settler's cabins, a mill, and a coal mine to a town with eleven general stores, two butcher shops, three bakeries, and two restaurants. Merchants had set up in tents to market their goods. Sailing vessels arrived from San Francisco with cargoes of clothing, mining tools, groceries, and other provisions. As of the 1st of July, an estimated 10,000 people were living in makeshift shelters on the shores of Bellingham Bay.[(2)] Neighboring Sehome was also being deluged with miners and a new wharf was being built for incoming boats. The restless throngs were waiting for the Fraser and Thompson Rivers to subside and hoping that very soon the trail leading to the mines would be ready for travel.

The San Francisco Herald announced that the stampede north was equal to the migration of prospectors from the Atlantic states to California in 1848, 49, and 50. As hundreds arrived at Bellingham Bay, hundreds of others were getting ready to make the return trip home. In July of 1858, the "Oregon" delivered 500 passengers, then left for San Francisco with 400 aboard. The "Santa Cruz" dropped off 300, then left with 100. After the "Pacific" unloaded 175 tons of freight and 150 passengers at Victoria on August 5, over 2000 crowded onto its decks to return to San Francisco. At Whatcom, 200 destitute miners took possession of the steamer "Panama" and threatened to navigate it back to San Francisco. It was unknown whether they carried through with their threats. However, there were those who sympathized with the miners since it was felt the company had prematurely brought them north. Provisions were scarce and expensive. At Whatcom, flour sold for $12.50 per hundred and the price was going up.

In July, 1500 had gone ashore at Victoria and Bellingham Bay. Whatcom was beginning to take on the appearance of a permanent settlement, with streets being laid out and buildings under construction. A brick building was even being built. Captain DeLacy sent news that he believed he had reached the summit of the mountains and that the remaining stretch would not be too difficult. Miners with pack animals were cautioned, though, not to start up the trail. As of August, the trail was still not completed. There had been some backtracking in an effort to find a more accessible route. Indian guides were along, but they, too, were faced with unfamiliar wilderness country. Some optimistic miners with pack horses started on the trail and ended up going a direction which finally led them to Fort Hope.

During the summer of 1858, the pioneer author Ezra Meeker was in Whatcom. He had taken some cows north from Steilacoom to sell milk and butter to the prospectors. While there, he observed a company of sixty men with heavy packs leave Whatcom to travel the trail. After several weeks of wandering through the mountains, they returned to the place from which they had departed. The first forty or fifty miles of the trail was well worn by men and animals who left but were then forced to return. Being unable to reach the mines was no doubt discouraging to those who had nurtured for so long the dream of striking it rich. To the weary, ailing, bug infested prospector, the thought of returning to home cooked meals and a feather bed must have seemed a welcome alternative.

When DeLacy reached a point which he believed to be the summit and then a short distance to open country, he actually had fifty or more miles of rugged mountains ahead of him. Meeker mentioned that DeLacy never did find a trail through the mountains and, as far as he knew, never did return to Whatcom. However, to keep hopes alive and prevent the collapse of the ongoing boom, DeLacy was kept in the mountains working on the trail which was rumored to be moving toward completion. (3)

In December of 1857, Governor Douglas of Vancouver Island exacted a mining license fee of those traveling to the Fraser River mines. In May of 1858, as Factor for the Hudson's Bay Company, Douglas issued a mandate requiring a license and sufferance fee for the transporting of goods up the Fraser River. Among other requirements, vessels could transport only those goods of the Hudson's Bay Company, and passengers going to the mines must have the license to mine. The mining license could be obtained only at Victoria, which served to direct traffic and business into that port. The license requirement and other opposing conditions led to the rapid decline of Whatcom's population:

> *In this way the Whatcom game was blocked, with or without a trail, and the population disappeared nearly as rapidly and more mysteriously than it had come, and the houses that had been built were left tenantless, the stakes that had been set were left to be swept away by the tides or to decay, and Whatcom for a time became only a memory to its once great population.* (4)

During the summer of 1858, an expedition traveled by canoe up the Skagit River in an attempt to intercept Captain DeLacy's trail which was to have crossed the upper Skagit. An Indian village was found near the "Takullam" fork entering from the north. This was about forty miles from the entrance to the Skagit. Near the mouth of this stream was the "Smalehu" Indian village. No white man was known to have ever traveled beyond this point. About five miles further, the party came to the "Sakumehu" branch of the river, flowing from the south. Here they noticed an island with three horses on it. The explorers were unable to determine where the horses had come from but an interesting piece of information provided a clue as to why they were there. It had been reported that a trail ran from this branch of the river to Lake Chelan across the summit. When asked about the trail, the Indians could not or would not share any information. [5] If there were actually an Indian trail which went over the mountains to Lake Chelan, it may have joined the trail reported by Chief Owhi and other Indians. The trail traveled north from the lake. Prospectors had recommended it as a possible short cut from Snoqualmie Pass to the Thompson River. Those who favored the trail estimated it to be only 300 miles from the pass to the river in Canada. [6]

The expedition continued up the Skagit as far as possible by canoe, which they estimated to be seventy miles from the mouth of the river. Here they met some Indians who were fishing. The Indians told them that there were immense rocks ahead and that no one, including Indians, had ever gone beyond that point. They were further advised that the mountains were impassible. With this information, the explorers decided to go no further and concluded that the Skagit River was not a feasible route to the Fraser River mines. This opinion would stand among prospectors for many years to come.

The Fraser River Route

In the discussion of 1860, a miner wrote regarding the Fraser River Route:

> *As to the Fraser River route, its character has been so often described that its whirling eddies, rapid currents and rocky shores, threatening death and destruction to all that may venture upon its uneven surface, are familiar to every reading man. . . .*

A miner could travel up the river on the ice but this would be a decision "... of the practicability of which the boldest cannot but be incredulous. . . ."

A letter received from the Fraser River, dated August 8, 1858, and printed in the August 20th Pioneer and Democrat advised that no one should undertake to travel the river beyond Fort Langley without the help of an Indian pilot. Some prospectors had traveled from Fort Langley to Fort Yale without a guide. They arrived three days late and were described as:

> *. . .worn down with fatigue, cursing the river as the worst in the world, when the truth of the matter was, the lack of knowledge how to manage, and the belief that they "knew it all" better than any Indian.*

Beginning at Fort Yale was a stretch of the river referred to as the "Canyon." There was actually a series of canyons which continued for forty miles, the first of which was eight miles long. Because of high water, it was almost impossible to travel the canyon above Fort Yale. As an alternate route to the upper Fraser and Thompson rivers, miners were now going by way of Harrison River and lake. Even at that, there were approximately 7,000 men concentrated on the fifteen-mile stretch of the river between Fort Hope and Fort Yale, but not more than 200 of them were making expenses. Many had become discouraged and were selling whatever they could to leave the country. Then there was the added threat of unfriendly Indians if miners should ever get beyond the canyon. The Indians had declared that no "Boston man" would be allowed to mine in their territory. According to the Vancouver Island Gazette, working conditions on the river improved greatly during August of 1858. As of the 14th, the river had dropped several feet and an estimated 10,000 miners were working near Fort Hope. Sixty pounds of gold had been brought down to Victoria and was now in the hands of Wells Fargo and Company. [7]

The estimated distance from the mouth of the Fraser River to Fort Langley was forty miles: from Fort Langley to Fort Hope, a distance of sixty miles; then a distance of fifteen miles from Fort Hope to Fort Yale.

Cady's Pass

The September 14, 1860, Pioneer and Democrat, announced that a low elevation pass had been discovered by Mr. Cady of the "Snohomish River." (The Snohomish River settlement called Cadyville later became known as Snohomish). The pass had been discovered eighteen months earlier and was expected to rival the Snoqualmie as a way to the mines. Cady, accompanied by others, had recently left to travel the new route. These were "hardy, courageous, and enterprising men." An earlier letter, dated March 19, enumerated several reasons why this pass should be given financial support:

> *Contrast the expense of freighting from the Sound by this route, with that of Oregon, and a large profit results to the merchants of our territory on the freight alone. Is it not then of deep importance to the farmers, millers, and merchants, and every class of producers in the Territory to put the financial shoulder to the wagon and push it along. . .*

Over $2,000 was subscribed from Port Townsend and Steilacoom residents toward the opening of a trail through the mountains by way of Cady's Pass. Builders ran out of money before completing the trail to the summit. E. F. Cady, E. C. Ferguson, and a Wenatchee Indian set out with a pack train to investigate mining activity east of the mountains and demonstrate the practicality of the route. As recalled by Ferguson, they reached the end of the recently built trail about five miles from the summit. They cut their own trail over the summit, then traveled down the (Little) Wenatchee River to Lake Wenatchee. They built a raft and worked their way across the lake, swimming the horses. From there it was open country to the Columbia River. They traveled up the Columbia, crossed the Entiat River, then at the foot of Lake Chelan met a man living with an Indian woman. The man ferried them across the lake in a canoe, again swimming the horses. They continued north to the mouth of the Similkameen where they crossed the Okanogan River, then traveled east to the Kettle River where only five claims appeared to be paying much. Cady and Ferguson were unimpressed with their findings, so the Cady Pass project which was to provide a route to the mines was subsequently dropped. Ferguson was quoted as saying, "We returned to Snohomish, but on account of the very limited extent of the diggings, the rush subsided and none else used our trail." [(8)]

Naches Pass

Although it appears to have not received as much publicity at the Snoqualmie Pass, the "Nah-ches" Pass was considered a suitable travel route. It was located south of Snoqualmie Pass and went from the headwaters of the White River on the west slope, to the headwaters of the Naches River east of the summit. During his historic trek through the northwest in 1853, Theodore Winthrop was guided over the Naches Pass by Loolowcan, son of the Yakima chief, Owhi. Winthrop and his young Indian guide left Fort Nisqually on August 14. While traveling toward the pass, they came to a section of road which had been opened by white settlers in 1850 and then more recently improved. He followed the road until the "slashings" made it too difficult to travel, then returned to the Indian trail which led across the summit. He continued his journey down the Naches Valley and on to the Dalles. [(9)]

A writer in the July 2, 1858, Pioneer and Democrat praised both the Snoqualmie and Naches Passes as travel routes for miners:

> *We regard the Snoqualmie and Nah-ches passes as at present the only "fixed facts" whereby adventurers can reach the goldfields east of the mountains, starting from and lying over the American soil, this side of the Columbia River. . . .*

HORSES, MULES, AND MINING

The horse was a friend and helper to the early prospector. It provided an indispensible means of transportation over the most difficult terrain. When pack animals were needed, mules were sometimes preferred because of their endurance and sure-footedness. Endowed with a tough hide and sturdy hoof, the mule was capable of scaling steep and rocky hillsides. Haulage for an average mule was around 200 pounds, although in some instances miners loaded over 300 pounds on the back of a large mule.

Mules were also used inside mines to haul supplies and pull ore cars. In some cases they were lowered in slings and stabled permanently underground but were brought out if they became ill. To house them, stables were cut out of the rock at the main level of the mine. One reason mules were preferred inside the mines was because they were considered to be less skittish, therefore, less prone to injury. While the pony has a tendency to rear its head when something touches its ears, the mule will usually duck. This would have been important to a miner working in close underground quarters.

The mule has been credited by some as being more intelligent than the horse. Enthusiastic owners have even suggested that they can count. When pulling ore cars, a mule could feel the tug between each car. If it were accustomed to pulling three cars and the miner attempted to add a fourth, the mule knew the difference and would balk. It could be fooled, though, by the miner cushioning the jolt of the additional car by slowly adding tension to the chain. Miners also claimed that a mule was capable of sensing movement or instability in the mine and would become agitated, thus warning them of an impending cave-in.

My long-time friend, Jim McDonough, who is well acquainted with mule behavior, tells a story which he says illustrates the mule's disposition. Jim knew a mule owner who, in the 1920's, drove his mule and wagon around the town of Marysville, Washington. The first time he came to the railroad

Joe Mills, a well-known prospector, with his pack animals next to the Tungsten mine dining hall in 1916. The Tungsten mine was located northwest of Loomis in the Pasayten Wilderness area.

tracks, the mule balked. It refused to cross the tracks. Studying the situation, the owner decided to try a little trick to see if he could divert the mule's attention away from the tracks. He found a small round stone which he used to tap the bottom of its front hoof, at the same time convincing it in a sympathetic tone that there was something wrong with the hoof. While periodically tapping on the hoof, he coaxed the mule forward a little bit at a time. By this means he was able to maneuver the mule across the tracks and continue on his way. In the future, he always carried a small round stone with him to use when he came to the railroad tracks.

Early prospectors learned to maximize an animal's strength for the purpose of transporting goods. One method referred to in the 1800's was "rawhiding." This apparently involved the use of a cow's rawhide in which goods were dragged along on the snow, usually downhill. Such an instance is mentioned in the September 5, 1895, Snohomish weekly, The Eye:

> *Preparations are being made to 'rawhide' ore down from the Kennedy Mines near Silverton as soon as the snow falls. By this method, a pony can transport half a ton or more at one time, while but 150 to 200 pounds could be packed on its back. The mines are about 2½ miles from the railroad and some of them contain high grade ore.*

It can be assumed that the horse trails leading from the mines were winding, narrow and steep, thus requiring the flexibility of the method of rawhiding as opposed to using rigid sleds or wagons.

Prospectors were advised to never leave their horses unattended and to use every means possible to protect them from being stolen. One recommendation in 1862 was that miners band together into fifteen or twenty so they could pay a person good wages to eat and sleep with their horses. Also, a miner should avoid leaving his horse with just anyone who claimed to have a ranch. A suggested alternative would be to shoot the animal. Then, at least, the owner would know where to find it. The sad truth is that horses had been entrusted to caretakers who had no intention of returning them to their rightful owners. In addressing this problem, the writer of a letter published in the June 14, 1862, Washington Statesman of Walla Walla listed the following suggestions:

Index packtrain in 1896. Index, located at the junction of the North and South Forks of the Skykomish, was a supply center and jumping-off place to other mining camps. To the north was Galena and Mineral City located in the Silver Creek District. An article in the Spokane Review, January 2, 1892, made mention of the great amount of mining activity in the Silver Creek District - "...the most active prospecting since 1859." In 1892, Mineral City was considered the principal town in the district, with twenty-four log cabins, one or two hotels and stores, a drug store, saloon, and blacksmith shop. Mineral City was the last camp to be visited before crossing the divide into Monte Cristo. (Photo courtesy of Snohomish County Museum)

. . . To judge the ranchero's honesty by the length of his spurs - the longer his spurs, the less his honesty. Secondly - never trust a man who always wears in addition to long spurs, his leggings and a flourishing silk handkerchief about his shoulders - that kind of ranchero is always ready to leave on short notice. Thirdly - when you ranch a horse, if the ranchero treats you to more than his fees in liquor, consider your horse bound to Cariboo at once. Fourthly - avoid leaving your horse with a man who carries more than one large six-shooter, and a bowie-knife, for they are the most dangerous of all.

The cost for transporting goods by pack animal was determined by distance and weight. In October of 1859, a packer's charge for freight from Walla Walla to Fort Colville was $11.00 per hundred. This figures to be 11¢ per pound for a trip that covered a distance of about 250 miles. Because of the time and difficulty in transporting to remote locations, the cost of goods often doubled or tripled. Flour sold for $18.00 per hundred at Fort Colville. At Steilacoom, 100 pounds wholesaled at $5.00. In January of 1860, merchants at Fort Colville had sold off their stock and were waiting for a new shipment from Walla Walla. Owners of pack trains were offered 20¢ per pound for a distance not to exceed 200 miles, but refused the offer. (The figure of 200 miles was a conservative estimate) They did not want to risk their animals during the worst weather of the year. In August of 1860, the following were other items being sold at Colville: bacon per pound - 37½ ¢ to 45¢; coffee, 40¢; sugar, 35¢ to 50¢; tea, $1.00 per pound; turnips, $3.50 per bushel; hay, $12.00 to $15.00 per ton. Laborers at Colville were scarce and wages high, about $3.00 per day.

Near Rock Creek in the fall of 1860, miners were paying 20¢ a pound for fresh beef and 50¢ a pound for bacon. A letter received from America City, dated November 26, suggested the need for a good

Pack train preparing to leave Glacier to travel to the Lone Jack mine in the Mount Baker District in 1900. "Supplying remote mines with food, tools, mining equipment and other items was a logistical nightmare in a region where trails were few or non-existent." (Jeffcott collection, courtesy of Center for P.N.W. Studies, W.W.U.)

overland road to help lower costs. The writer also expressed his feelings about the use of pack animals for transporting supplies:

> *What we want now is a good wagon road from the Dalles to Kettle River - one upon which we can put our six-mule teams and freight our goods at a better advantage. Packing is the last resort of transportation in any country. It is done at great expense and is, at the best, a very hard slavish work.* (10)

The writer then digressed about "Old Abe" being elected the next president and the fact that the snow would soon close off land communication over the hills. They would then be left to themselves. He commented further that if there were stations along the road, it would be nice country for winter travel. The snow was not too deep for sleighing.

When provisions had to be packed to destinations on the backs of hikers, the cost really accelerated. In the Salmon River area in March of 1861, supplies had been packed twenty-five miles to reach miners in the mountains. By this method, ordinary freight ran 40¢ to 50¢ per pound, flour - $1.00 per pound, long handled shovels - $16.00 apiece, and camp kettles at from $8.00 to $10.00 each. Other items in demand by miners were syrup, vinegar, heavy flannel clothing, soap, towels, and anti-scorbuties (a remedy for prevention of scurvy).

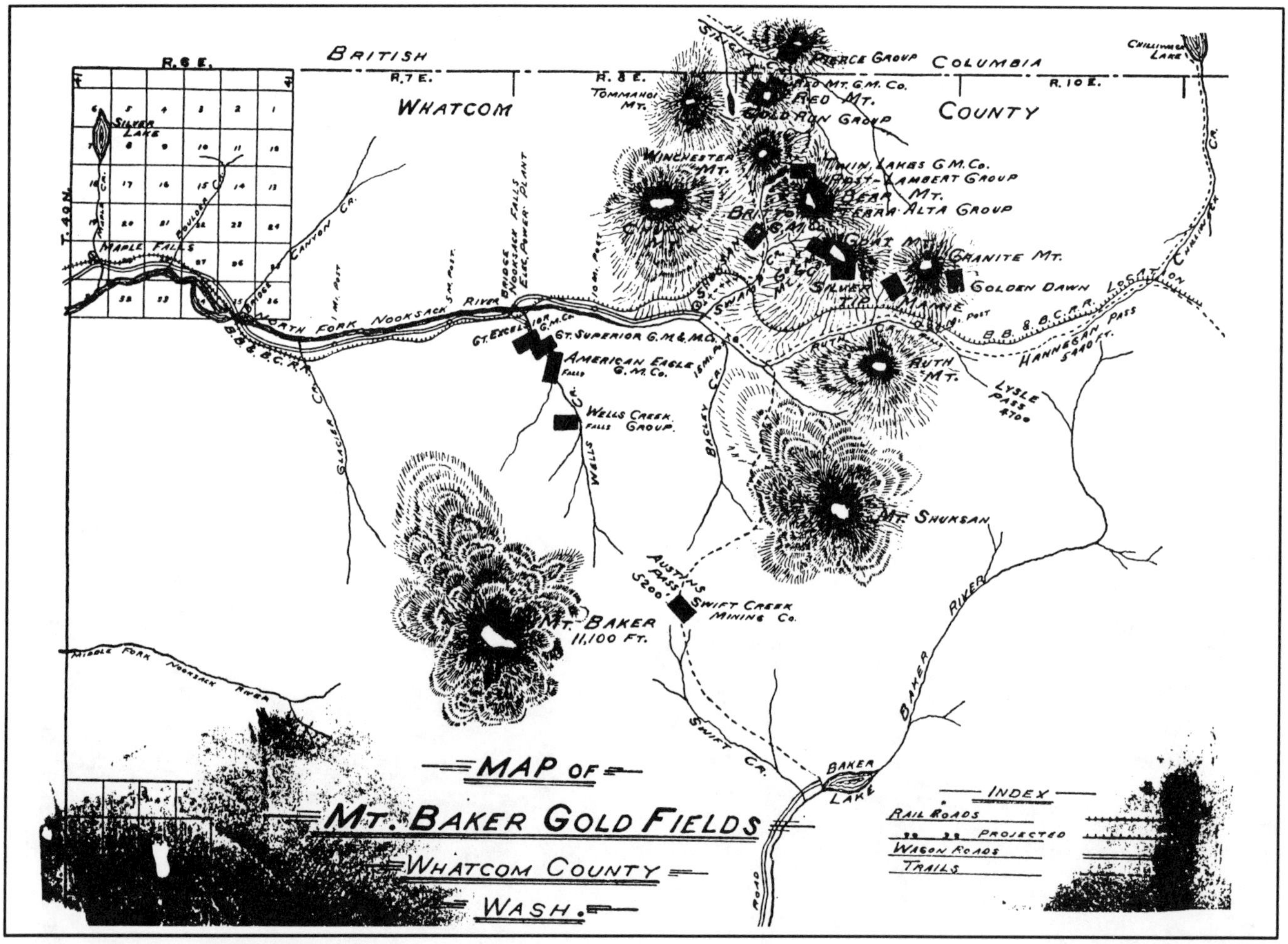

Old U.S.M.S. Map of the Mount Baker Mining District in Whatcom County. The Lone Jack became the major producer of gold in the district. Many unusual specimens of free gold were taken from the mine. The mine was located on the east slope of Bear Mountain at an elevation of about 5,000 feet. The original discovery was made in the summer of 1897. Because of deep snow during the winter months (sometimes 20 feet deep), the mine usually operated only 5 or 6 months of the year. Snowslides, too, were a constant threat. It is estimated that during the years of operation from 1902 to 1924, the mine produced about $550,000 in gold. (Information from W.S. Moen, Bulletin 57, 1969).

During the 1850's, there were only "estimated" distances from place to place. One might suspect that haulers or guides were eager to take the benefit of doubt when contracting for a certain destination. According to an estimate given in the July 31, 1858, American Railroad Journal, the distance from Fort Walla Walla to Fort Colville was about 400 miles when traveling the Columbia River route. By following the Columbia, Fort Colville was believed to be 700 miles from the entrance of the river. A letter sent from the Dalles indicated that by cutting directly across the country, it took about five days to travel with mules from the Dalles to the Similkameen in the Okanogan. If they were actually able to cover the distance in that amount of time, they would have had to really push it. With pack animals, it took five days to travel from Fort Colville to the Similkameen diggings and only a day and a half by express. Today, it takes from two to three hours by car.

Those carrying the mail used faster horses and prided themselves in riding the distance between posts in the shortest possible time. The express left Rock Creek on Monday, August 13, 1860, and arrived in Seattle six days and five hours later. If the total distance covered had been 360 miles, then the rider would probably have traveled between fifty to sixty miles per day. This is remarkable considering the fact that much of the trail was over steep hills and through numerous bends. The number of different horses ridden during the six days is a matter of guesswork.

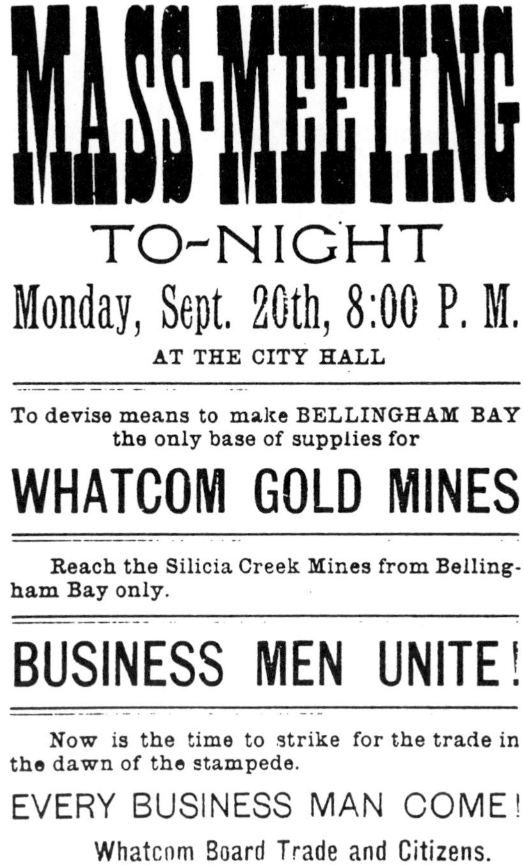

1897 Handbill issued following important discoveries by Post and Lambert on Silicia Creek, known today as Silesia Creek, a few miles east of the Lone Jack and Boundary Red Mountain mines in the Mount Baker District. This flier illustrates the eagerness of early merchants to capitalize on miners traveling through the country. (Jeffcott Collection, courtesy of Center for P.N.W. Studies, W.W.U.)

Twenty-four horse team pulling a boiler to the Copper World mine on Palmer Mountain above Loomis, Washington. The scene gives added meaning to the term "horsepower" used as a measuring unit for power or rate of work. (Photo courtesy of the Okanogan County Historical Society)

THE SWAUK MINING DISTRICT

In the Cascade foothills south of Swauk Pass is the Swauk Mining District and the historic town of Liberty. Along Swauk Creek near Highway 97, mounds of tailings remain where a bucket-line dredge operated in the 1920's. Near the confluence of the Swauk and Williams Creeks, a road turns east from Highway 97 to the townsite of Liberty. The first discovery of gold on Swauk Creek was in 1868. In that year, a deaf-mute who had been hunting went to the creek to get a drink of water. To his surprise, he found a nugget lying on bedrock. As soon as the word got out, miners rushed to the creek to search for gold. (11) Since that time, the area has been mined almost continuously and has been one of the major gold producing areas within the state.

In the early 1890's there was a resurgence of mining activity in the district. New discoveries of both lode and placer deposits were being made. It became understood, though, that the gold could be recovered only by a great amount of time and labor. Much of the effort of the miners was devoted to working ancient river channels which were in most cases covered by many feet of overburden. They endured formidable odds to dig deep shafts to bedrock and then drift tunnels along the bottom, following the course of the gold. Wherever the old channel was struck on either Williams or Swauk Creek, rich deposits were found with good pay. Many of the drift tunnels ran for hundreds of feet and at least two were reported to be over a half mile long. If an overhead view of the underground diggings were possible in our time, one would see a network of old tunnels leading in many different directions.

Huge nuggets of many ounces were found on bedrock. In the late 1880's, one claim yielded three large nuggets which ranged in value from $200 to $400. In the spring of 1891, John Black found a nugget at the Old Discovery claim worth $726. The

Information sign at Liberty.

claim was sold to the Green Tree Mining Company of Tacoma for $6,000 after reportedly taking out $50,000 worth of gold. [12] A champion sized nugget was found on a high bar near Baker Creek. It was valued at $1,004. At that time, gold from Baker Creek was sold for about $13.50 per ounce. [13] In today's market, as a collector's item, the nugget could be valued at well over $30,000. Since the miners were more concerned with high-grading materials close the bedrock, lots of paydirt was discarded into tailings. Some the these tailings piles have over the years been reworked with good results.

Areas of placer ground along Williams and Swauk Creeks are still being mined today and handsome nuggets are being found. Fashioned by dynamic forces in prehistoric times, many of the nuggets are found in the form of intrinsic wire configurations. While some appear as a ball of enmeshed wire, others display distinct patterns of beauty. By the slightest effort of imagination, one can see in many of them the resemblance of familiar objects such as flowers and animals. Thus are the surprises of Liberty gold - a fisherman's fly, wings of an eagle, or a Christmas tree. The common expression is "Oh, that looks like. . . ."

In addition to the gold which has been found in local placer deposits, lode mines in the Liberty area have over the years produced rare specimens of leaf and wire gold. Discoveries of gold-bearing veins date back to 1881. In 1887, two prospectors found a pocket of gold on the east bank of Swauk Creek, opposite the mouth of Baker Creek. Years later, an unusual specimen was found in the same vicinity. It was a large conglomerate of wire gold which had the appearance of a basket complete with handle. Old timers still talk about the nugget that was called the "Hanging Basket."

In 1891, an unusual auriferous formation was being mined at a "tip of the summit" location. The

Scenic view of Liberty country with a gold recovery plant in operation. A trommel is being used to separate gold-bearing gravel. Worthless rock is discharged by elevator onto a tailings pile. These hills have been heavily prospected since the 1800's.

vein, as it was called, was only four inches wide and was composed of a type of clay but was rich with gold. It lay near the surface, not more than six feet deep. During the winter, when the ground was frozen, boiling water was poured over the material to help break it up. It was then pulverized in a mortar, retorted, and shipped. (14) In 1878, a newspaper article mentioned a "blue clay" pay dirt which was being mined in the Swauk district. The clay contained coarse quartz and gravel which were mixed together with gold. (15)

Early miners also gleaned gold from an unusual vein material called "bird's eye" quartz. Fine grains of gold occurred within the quartz or next to fragments of black shale. The small quartz crystals often projected into open spaces. In some cases, encrustations of leaf or wire gold were found on the surface of the quartz. A specimen taken from one of the mines contained perfect octagonal crystals of gold attached to the end of quartz crystals.

Miners were always looking for a pocket which might contain gold, possibly many ounces of it. Around 1953, a fabulous strike was made by Clarence Jordin Sr. when he discovered a rich pocket in the Ace of Diamonds mine. It yielded a 134 pound mass of gold. Thousands of ounces of gold were harvested from the famous Ace of Diamonds.

A unique collection of local wire gold comprising all the letters of the alphabet was gathered in the 1940's and donated to the State Historical Museum at Tacoma, Washington. Because of its value in commemorating American history, Liberty has been registered as a National Historic Landmark. At the townsite, a replica arrastra with an information sign is on exhibit.

The color-penciled map with legend on the back cover describes the Swauk and Peshastin (Blewett) districts at the turn of the century.

Miner's pick and wire gold typical of the type of gold found near Liberty, Washington. (Photo from Washington Division of Geology and Earth Resources files)

CHAIN-BUCKET DREDGE

The biggest of all mining devices, the chain-bucket dredge (also called the bucket-line or bucket-ladder) floated on a scow in the river or sometimes near a river or stream where it navigated in its own self-made pond. Because of its size, it was nicknamed "the mountain boat." A version of the dredge was introduced to California from New Zealand in the late 1890's. During the following decades, dredges were installed in all of the major gold fields of the American Continent.

The concept of the dredge evolved from a method used in New Zealand to recover gold from riverbottom placers. Miners used a device called the "spoon," which was a bag woven or riveted around a metal frame and attached to the end of a long pole. Using weights to hold it down, miners dragged it along the bottom of the river to collect gold-bearing gravel. When it was partially or nearly full, it was brought to the surface and emptied into a rocker which was situated on an adjoining boat. The scheme was improved upon when they placed a platform onto three canoes and worked from the platform. Later, they added the mechanical features of water power and crab-winches to operate the spoon.

Chain-bucket dredge which operated on Swauk Creek in the 1920's. The dredge was the largest of its kind in Washington State. Over $35,000 in gold was recovered during the years from 1926 to 1929. (Photo from Washington Division of Geology and Earth Resources files)

In 1870, a man by the name of Mr. Ward designed the bucket and ladder system which was run by a current wheel and later by steam power. [16] Improvements and modifications were made but the chain and bucket elevator continued to be the distinctive feature of the dredge. A series of steel buckets were attached to a chain which ran along on rollers and sprocket-wheels or tumblers that were located at each end. The upper tumbler drove the bucket-line. This system was considered desirable because of the huge amount of auriferous material that could be delivered from deep subterranean placer deposits. There were buckets of different sizes, ranging from smaller ones having a capacity of three cubic feet to larger ones of eighteen cubic feet. The gold recovery rate for the dredge has been estimated to be from 50% to 60% or possibly higher depending on its gold-saving plant.

The gold-saving plants in the dredges were not as standardized as the other components. Common fixtures, though, were the trommel screen and sluice. Typically, materials lifted by the buckets were emptied into a hopper, then ran through a revolving trommel to be classified. High velocity jets of water were introduced inside the trommel to wash the gravel. Holes in the trommel screen varied in size, from 3/8 inch at the receiving end to five or six inches diameter at the discharge end. The trommel could be three to four feet in diameter and from twenty to thirty feet long, depending upon the size of the buckets. Sometimes the trommel consisted of two screens, one inside the other. Material from the inside screen went onto a sluice while material from the outside screen was distributed over gold-saving tables. Larger materials, from fist-size to heavy boulders, which passed through the end of the trommel were dropped onto a stacker belt to be conveyed to the rear of the dredge. In those locations where the bucket-lines once operated, tailings piles have provided good hunting grounds for nuggets which were classified as too large and consequently were discharged with worthless rock.

DIVERTING WATER FOR MINING

During early mining activities, innovation often meant the difference between gold or no gold. Various creative means were used to divert and control water for the purpose of mining. With great toil, ditches, flumes, canals, wing dams, and cofferdams were built. In some cases, streams were even rerouted to gain access to the gold.

Wing dams were built on many of the rivers of the northwest before and after the turn of the century. Such a dam was built at Rich Bar on the Similkameen around 1860. The miners were forced to discontinue work when a flood destroyed the dam.

In the 1930's, Samuel Gjerde, a gun shop owner who lived near the Similkameen River, undertook to build a wing dam where he believed gold would be found on bedrock. He used a hand winch to maneuver large boulders into place. Weeds and dirt were then stuffed into the cracks to hold back the water. With the assistance of an engineer, he built a suction dredge powered by a Model A engine. It was designed with a small nozzle, ideal for extracting gold from bedrock crevices. Soon after they began working, they discovered the overburden was much deeper than expected and bedrock lay beyond the reach of the dredge. A hidden reef ran across the river. This was a big disappointment. As an alternative approach, they decided to bring in a backhoe. They dug a hole ten feet deep and never did hit bedrock, but still recovered over $1,000 in gold. While working below the dam, Art Gjerde, Samuel's son, was surprised to find grain-sized nuggets in an unusual place - on a false bedrock of compacted sand. As soon as water hit the sand, it disintegrated.

Another method used by early miners and especially the Chinese was "ground sluicing." This involved the cutting of a trench through a river bar.

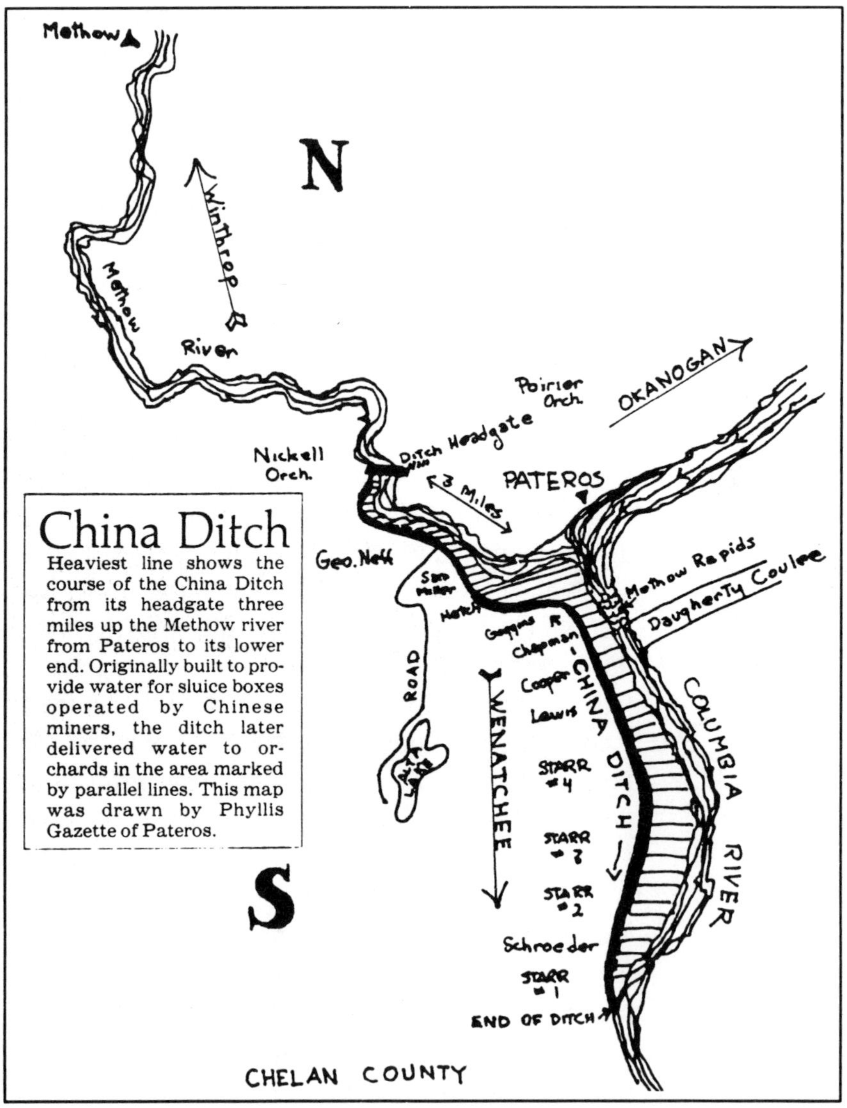

Map of China Ditch drawn by Phyllis Gazette of Pateros and published in the Okanogan County Heritage, Vol. 24, No. 4, Fall, 1986. A portion of the famous ditch still remains along the Columbia River below the town of Pateros. The canal was built by the Chinese sometime between the 1860's and 1880's to convey water from a headgate three miles upriver on the Methow to their workings on the Columbia. Where it was necessary, they whipsawed logs into lumber to build sections of the flume. When the Chinese discontinued their work below the confluence of the Methow and Columbia, the ditch was taken over by local farmers to irrigate orchards. The ditch represents a phenomenal undertaking by early Chinese miners.

The wing dam built by Samuel Gjerde on the Similkameen River in the 1930's. Wing dams were usually constructed in areas where the bedrock lay close to the surface. The purpose of the dam was to divert water away from a downstream area where miners hoped to find lots of gold. Out of the path of the strong current, they were free to work the riverbottom. Miners methodically moved boulders and debris in order to cover every inch of the bedrock. Occasionally, they found substantial amounts of gold on an impervious layer of clay called "false bedrock."

Hydraulic mining operation on Mary Ann Creek in northcentral Washington in 1906. The Allen Placer Company built the flume which extended over 1½ miles from Tippy Lake in Canada. A powerful jet of water washed auriferous gravels from the hillside into a long sluice. A special spout and nozzle mechanism, called the hydraulic "Giant," was used to propel a jet of water. Pressure at the nozzle tip was largely determined by the elevation of the water source and volume of flow. Gold was first discovered on Mary Ann Creek in 1888. Reportedly, $40,000 in gold was taken from the creek during the turn of the century. (Photo courtesy of Molson Museum)

Water and gold-bearing gravel were channeled over a sluice box situated at the lower end of the trench. Miners either shoveled pay dirt or caved it into the trench. Sometimes a head gate was installed so the water flow could be regulated or stopped completely. When it came time to clean up, the miners cut off the water, then used a pan, rocker, or sluice to separate gold from the concentrate deposited on bedrock and in the riffle box. They were always careful to clean any crevices or cracks on bedrock where gold might have settled out. By constructing a series of these races or trenches, miners were able to recover most of the gold from an entire bar.

The author's sons, Darren and Brad (wearing raincoats), working with friends on the Forty Mile River in Alaska. They had built a small wing dam to channel water over a sluice box. This was a method commonly used by prospectors in the 1800's, only on a larger scale.

RECREATIONAL DREDGING

The suction dredge performs similarly to a household vacuum cleaner, except water is displaced instead of air. Suction at the nozzle is created by a high velocity jet of water which enters the hose chamber at an angle. Gold-bearing gravel is vacuumed from the riverbottom and conveyed upward through the flex hose and then over a sluice where the gold settles out. The advantage of the suction dredge is that it can recover gold from normally inaccessible deposits at the riverbottom.

Most of the heavier gold found during my years of amateur prospecting was on bedrock, either at the edge or in the river. Nuggets have been found in unbelievable locations at the bottom of the river, lodged deep within crevices and sometimes wedged so tight that one would wonder how they ever got there. One of the best spots I ever found while dredging was around a section of elevated bedrock about fifteen feet long and four feet wide. It was in the main channel of the river and ran crosswise to the current. At the downstream side it dropped about three feet to the main bedrock floor. At one end of the bedrock section there was a short crevice with a pocket at the lower end. When the loosepack was scooped away and the bottom of the pocket exposed, there were several small nuggets just waiting to be picked up.

The downstream surface of the bedrock was broken into layers with seams running from top to bottom. I moved around the base of the bedrock, checking each little crack and crevice. Wherever it looked as though a nugget might be lodged, a screwdriver was used to carefully pry into and break apart the rock. I was surprised to find, more often than not, that a nugget came falling out. Within an hour, a nice collection of small nuggets were collected from this

Suction dredge built in the 1930's by Samuel Gjerde to work the Similkameen River above Oroville, Washington. As an experimental model, the Gjerde dredge was a forerunner of the modern recreational dredge used on many of the rivers of the northwest. The dredge was powered by a Model A engine. The winch at the rear was used by Mr. Gjerde to move rocks while building the dam to cordon off the river. (Photo courtesy of Art Gjerde)

one short stretch of bedrock.

Under water, small nuggets can be seen from several yards away and can be picked up with the fingers. Color and size are greatly amplified. A nugget which may be only 1/4 inch long appears to be at least twice that size. Each nugget has its own distinct features of color, size, and shape. It has been my experience to have never found dull or tarnished gold under water. Shining from the river floor, even gold fines (small particles) glitter with an unmistakable radiance. Thus, the modern prospector can share the same kind of excitement as was experienced by the very first miners. As in the early days, especially prized are those nuggets which contain quartz matrix. The quartz makes them difficult to solder for jewelry, but they do make beautiful showpieces anyway.

The Hookah (surface air supply) apparatus is an indispensable extension of the underwater dredger. This is his air support system. It has many advantages over the popular SCUBA (self contained underwater breathing apparatus) used for sports diving. The source of air is an oil free compressor which is driven by the dredge engine. An air line runs from the compressor to a reserve tank and then to the diver where a regulator supplies air on demand. The diver has an unlimited air supply and there are no tanks to be refilled.

There are many cautions about which the diver should be aware. The compressor and other components should be of the type sold specifically for the Hookah system. A substitute or makeshift compressor may compress deadly oil fumes into the air supply. Motor exhaust fumes, if conveyed to the diver, is also very dangerous. While diving, there are numerous safety rules which should be observed. Undermining a large boulder, causing it to roll over, can trap or injure the diver. It's a good idea for any dredger to know the same basic information required of those who SCUBA dive, although some of the hazards may be different.

The idea of using underwater devices to recover gold from the riverbottom was being explored as far back as the 1800's. An interesting story appeared in the September 9, 1893, Everett News under the caption, DIVING FOR GOLD DUST:

> *Captain A.O. Benjamin, the diver, returned Thursday night from a prospecting tour on Ruby Creek, about 85 miles above Hamilton, Wash. His principle object, as published in the Telegraph at the time he left, was to ascertain whether or not there was placer gold in the bottom of the creek, and to that end he explored the stream as far up as the old Nip-and-Tuck claim. Captain Benjamin had his diving suit with him and made a hundred or more dives to the bottom of the creek, and at no time did he fail to find gold. It is his opinion that there is paying quantities of the precious metal all along Ruby Creek from its mouth to the Nip-and-Tuck claim.*
>
> *There is some talk on the part of those owning mining claims of changing the course of the river and in several places wing-dams are being put in so as to drain the creek for a distance of several hundred yards. Captain Benjamin has for two years past been working on a machine which he thinks will, when completed, take the gold from the river bed without the expense of changing the current.*
>
> *In time, Captain Benjamin thinks a wagon road or railroad will be built into the Ruby Creek country and that there is sufficient ore there to warrant the construction of one or the other. He says a wagon road can be constructed from Goddell's Landing up the Skagit river to the gold fields for about $30,000.*

The May 4, 1894, issue of the The Eye announced that an experimental gold dredge was being worked on the Fraser River. Other than using a pump, few details are offered as to the function of the machine:

> *The trial of a new gold dredger for working sand and gravel in the bed of the Fraser River, proved a success at Yale last week. Working in twelve feet of water, gold gravel was pumped up from the bed of the Fraser River which yielded nineteen and one-half*

ounces of gold in a thirty-six hours' run. Stones weighing twenty-five and thirty pounds were brought up by the pump, and the whole machinery worked perfectly.

In 1896, a gold dredge was brought all the way from Wenatchee to Lake Osoyoos on the steamer Oro to work the bottom of the Similkameen River. The dredge was equipped with a centrifugal pump, sluice boxes, and a system for moving large rock. (17) According to a long-time resident who knows the history of the river, an eyebolt anchored in bedrock with a tie rod was used to winch the dredge upriver. An eyebolt, believed to be the same, has been found just below the railroad trestle about three miles above Oroville. It can be seen only during low water.

THE SLUICE

During the 1800's, miners often did not have access to metal for the construction of sluice parts. Therefore, they had to improvise with whatever they could put together from nature. One of the most available materials was common rocks. These were placed on the bottom of the sluice as riffles. Okanogan Smith was to have used this method of sluicing and probably picked up the idea while in California. Another early method was to place round chunks of wood in the trough.

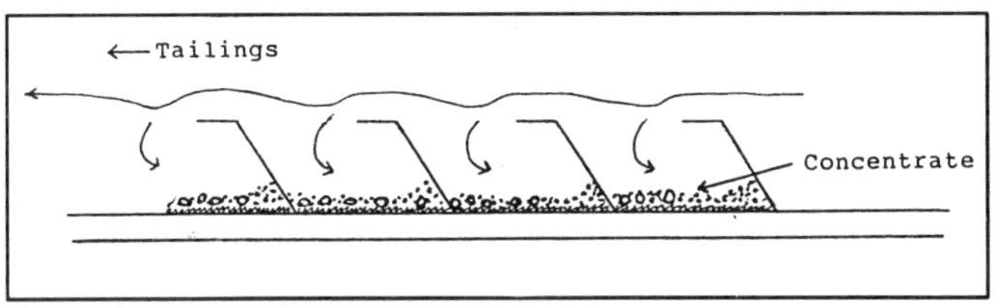

Hungarian riffles.

Today, riffles are made in a variety of styles from steel, but the principle is the same - creating a riffle action which allows the gold to settle to the bottom and lighter materials to run over the top onto a tailings pile. Popular among recreational dredgers is the removable Hungarian type riffle grill. Constructed of heavy sheet metal, each riffle is tapered at a slight angle in the direction of the flow of water, with a short horizontal surface. The riffles create a rolling action of the water which continually agitates the trapped material. This agitation process allows the gold to filter downward to the bottom. A layer of miner's or grass carpet is usually placed beneath the riffle grills to catch and hold fine gold.

In the 1930's, Bill Bleifuss and his friend, Oscar

Bill Bleifuss sluicing below where Baker Creek flows into Swauk Creek in the 1930's.

Knutson, were prospecting on Swauk Creek. An old-timer who knew the history of the area told them about some waste dumps that went back to the early days and still had gold in them. The first miners had tunneled into the high channels and gleaned only the heavier nuggets from the surface of the bedrock. Since they were mainly interested in working the bedrock, they did not bother washing the discarded material that had been carried by wheelbarrow to dump sites near the creek. Skeptical of finding anything worthwhile, Bleifuss and his partner tested the ground using gold pans. They got excited when they found two nice nuggets. This convinced them that the old-timer was right and that they should go ahead and set up a sluice. When the sluice was complete, it consisted of a series of seven boxes, each twelve feet long. The last one served as a riffle box. This gave the length needed to work several of the waste dumps along the creek.

After recovering several ounces of nuggets from the waste dumps, Bill and his friend started digging a tunnel on bedrock. One morning after breakfast, Knutson went out to turn on the water to the sluice and get ready to work. Bill followed a little while later. As of yet they had not opened the head gate so Bill decided to check out the tailings. When he got there, he looked down and there it was - a huge nugget right on top of the tailings pile, clean and shiny. What a surprise! They were amazed that anything that big could ever have made it over the riffles. Bill figured that it had rolled over the riffles in a ball of clay, then was uncovered after being washed for a time. It weighed 3¼ ounces. Those were hard times, so the men decided they had to sell the nugget. They got $75.00 for it.

Another even larger nugget was taken from the tunnel that Bleifuss and Knutson had dug. It weighed a whopping 4½ ounces. Bill has said,

Oscar Knutson, the partner of Bill Bleifuss, holding the 3¼ ounce nugget in his mouth to show its size. What a mouth full!

"Truth is stranger than fiction and this is a true story." A month or so after Bill and his partner had left the tunnel which was dug in about fifty feet, a prospector who had been poking around in the district went into it and sat down to light a cigarette. When he lit the match, he saw the reflection of a shiny object on the sidewall. It had the luster of gold. He went to check it out and discovered that it was gold - the exposed portion of a large nugget. The bedrock was straight up and down except for a hole where the nugget had been lodged and partially covered with clay. He dug it out and was amazed when its full size was revealed. Bill was not there at the time, but the story was told to him by the man who purchased it, a Mr. O'Brien. Bill was obviously disappointed that he had missed finding the nugget.

Other Amazing Stories:

Amazing stories are told about the finding of gold. The story is told that in the 1960's, a young man was playing with his dog, Charcoal, above Peshastin Creek in the Blewett District. The boy threw a rock for the dog to recover. When the dog returned, it had in its mouth a rock that looked like it had gold in it. Upon further examination, it turned out to be a chunk of quartz laced with copious amounts of the precious metal. The boy was offered a year's tuition in college for the unusual piece but turned it down. Word has it that the nugget is still somewhere in the Wenatchee vicinity and in the possession of its original owner.

Another unusual discovery of gold was in the 1920's. A farm lady who lived on Nine Mile Road above Oroville, Washington, was cleaning the gizzard of a goose when she discovered 3 small pea-sized nuggets. The goose had picked them up somewhere on the farm. The worth of these nuggets was more in the manner of which they were found than the weight of the gold. Three nuggets from the gizzard of a goose! What a story! The prized nuggets were placed in a cabinet to be brought out and shown when friends or visitors stopped by. One day they came up missing. It was suspected that a farm worker passing through had helped himself to the nuggets, worth a few dollars. Apparently, the finding of the nuggets in the gizzard above Oroville was not the first time that this had happened. The Everett Times in 1894 mentioned the practice of thrifty housewives examining the crops and gizzards of slaughtered water and land fowl. ". . . many a parlor mantel in Snohomish, Skagit, and Whatcom counties is ornamented with little vials of fine gold dust thus gathered."

Look at this! A rock with gold in it!

THE ROCKER

As gold was being discovered at various locations in the northwest, miners were always ready to move on to something better. In order to keep mobile, they had to rely on those tools which were light enough to be carried by foot or on horseback. One of the most popular portable tools used in the 1800's was the rocker, also called the cradle. It was lightweight, yet could process several times more material in a day than could be run through a pan.

The design of the rocker was universally the same. A removable receiving hopper was on top. At the bottom of the hopper was a sheet of metal with perforated holes. The holes allowed smaller sized materials and gold to filter through to an apron and sluice located below. Sometimes miners added an amalgamation plate which assisted in the recovery of extremely fine gold such as was found on the beaches near Nome, Alaska.

Usually two men operated the device. While one shoveled gold-bearing material into the hopper, the other poured water over it. A gallon bucket attached to a stick was frequently used as a dipper. One or the other of the workers kept the rocker moving continuously from side to side. Periodically, the hopper was removed, emptied, and refilled with pay dirt. Where water was in short supply, they operated over a wooden holding tank so the water could be recycled.

Another device used popularly during the 1800's was the long tom. Like the rocker, the long tom had a classifier for separating the gold. The design consisted of a trough with an inclined end (classifier) which had holes in it. Pitchforks were used to purge larger rock from the trough while smaller materials and gold filtered through the holes and onto a sluice. Sometimes water was diverted from a stream to provide a continuous supply for washing pay dirt.

Miners with their rocker on Skookum Creek in Whatcom County, 1901. (Photo from Washington Division of Geology and Earth Resources files)

The "Rocker"

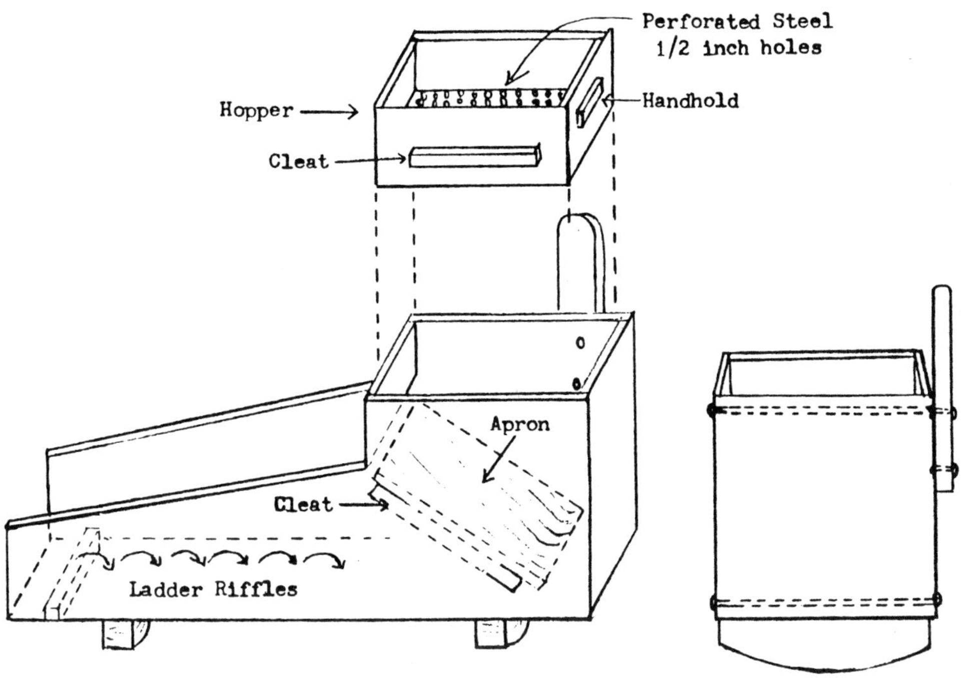

The "Rocker" was a very effective tool for recovering gold. Two men could wash from three to five yards of material in a ten-hour day, at least twice as much as by panning. Coarse gold was caught on the apron while finer particles were trapped in the riffles at the bottom. The length of a rocker varied from 24" to 60".

The "Long Tom"

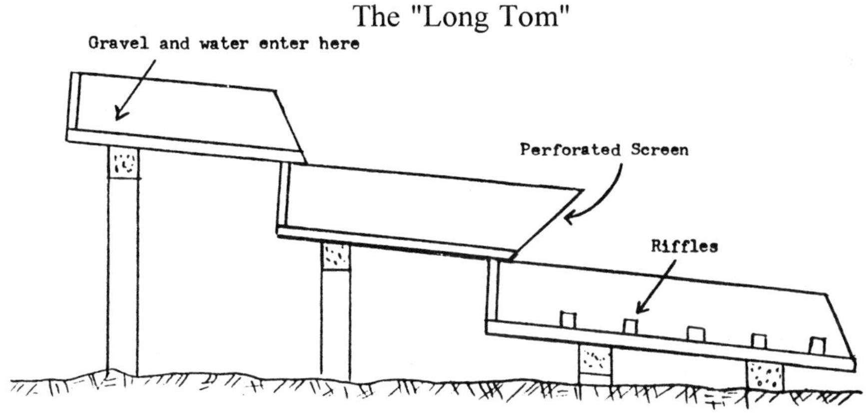

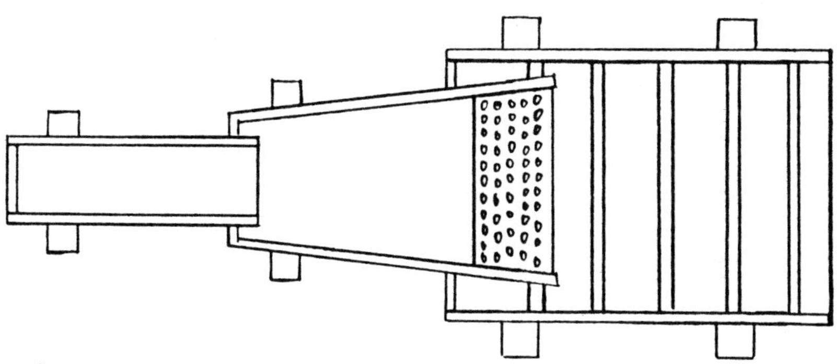

The "Long Tom" was a small-scale placer mining apparatus. It required flowing water to wash auriferous gravel. The perforated screen prevented coarse material from going onto the riffle-box. Hard working miners could run from four to six yards of material through a long tom in a ten-hour day.

BEWARE OF THE RATTLESNAKE

Some gold-producing areas in Washington State are ideal habitat for the rattlesnake, such as along the Similkameen River. In this terrain, one needs to be alert to the presence of these creatures and exercise the cautions necessary to protect oneself from a lethal bite. High boots are a must. One needs to be careful when moving rocks or other objects which provide cover from the sun. I've observed that rattlers like to come out in the evening when the hot sun is no longer overhead. They are attracted to stream locations where they feed on mice, birds, and other small animals.

While on the subject of the rattlesnake, there is another kind of subtle character that is worth mentioning and of which one may be advised to be cautious. He walks on two legs and lays in wait for uninformed prey to come near, ready to strike at the most opportune time. Shrouded by innocent looks and smooth talk, he's ready to tell of the riches to be had by investing in some claim, prefaced by the supposed fact that fabulous amounts of gold have already been gleaned. And so it goes. The unscrupulous culprit pockets the hard earned money and the vulnerable get bit. So, if one is really interested in purchasing any claims, leases, or anything else related to mining - be alert! It's better to be cautious than to get bit.

*Watch out! There's a rattlesnake! The large rattlesnake had just crossed the Similkameen River in front of the author's dredge. Such handling of snakes is **not** recommended.*

During the gold rush days, mining personnel were warned to be aware of the swindle called "mine salting." There were numerous devices used to deceive investors. One method of deceit was:

> *. . . to inject chloride of gold in the sacks containing samples and this can be done in an instant while the back of the victim is turned. Another way is to carry a common bulb under the arm with a tube attached running down the arm, to the end of which is attached a quill loaded with fine gold which can be blown into the samples, even when the victim is looking on. Changing the samples is another method which has many times been worked successfully. Mixing samples is oftentimes employed.*

Mixing small nuggets with a chew of tobacco was another means of deceit:

> *In placer mining the man who is showing the property is generally loaded with small nuggets, and while he is busily engaged he naturally spits (tobacco) in the pan, and the nuggets will be found there when the gravel is panned out.*

A known excuse for spitting in the pan was to bring good luck. In this case, the good luck was on the side of the swindler.

A nineteenth century mine expert went so far as to say:

> *For my part, when I am taking samples I would not permit my brother in the mine with me, even if he were a minister. Samples taken from a mine should always be pulverized and quartered down on the spot by the expert himself, the last quarter put in a double paper bag and sealed, which will in most cases eliminate the chances of its being tampered with. These samples, however, should never be out of the sight of the expert.* [18]

KEEPING LAW AND ORDER

When news of new discoveries were made known, hungry and eager prospectors rushed to the new El Dorados to compete for the best diggings. Given the primitive conditions and limited resources with which miners had to work, it is understandable that tolerance was not one of their leading virtues. To protect ones property was considered of great importance and sometimes even a matter of survival. Then, there were those within the camps who really believed they had traveled beyond the reach of the law and that justice was a matter of their own preference. These factors, combined with the problem of excessive drinking, created a volatile environment. Boundary disputes and accusations of high grading were all too often settled from the smoking barrel of a six-shooter.

To maintain a semblance of law and order, miners in some districts organized themselves into confederations to legislate rules and regulations. In September of 1878, miners held a meeting at the discovery claim of J.T. Cotton and M.R. Barrett on the Sultan River to draw up a code of laws for the district. About fifteen miners attended the meeting. The laws pertained to the whole of the Sultan river valley, from its source in the mountains to its mouth.

Miners in the Oro Fino (Orofino) District on the 5th of January, 1861, held a meeting to draw up mining laws. Excluding the additional resolutions enacted at later meetings, the laws as recorded and published in the December 28, 1861, Washington Standard of Olympia were as follows:

Bounds of Oro Fino District: Commencing at the mouth of Quartz Creek, and running up Oro Fino Creek five miles above Canal Gulch, including all the tributaries of said Oro Fino Creek within the aforesaid bounds.

Sec. 1. All American citizens may locate and hold claims.

Sec. 2. Each mining claim shall consist of 150 feet, running up or down the gulch or stream upon which it may be located, and extending from bluff to bluff on said gulch or stream unless said bluffs be more than two hundred and fifty feet from each other.

Sec. 3. No claim shall be considered forfeited from the first of December to the first of June on Oro Fino Creek, nor on the tributaries thereof from the first of December until the first of April.

Sec. 4. All claims shall be considered forfeited from the first of June within fifteen days from said dates if not worked.

Sec. 5. All claims shall be worked at least one day in seven, after the expiration of the said fifteen days; otherwise they are considered forfeited.

Sec. 6. Every person shall have the privilege of a drain race through any claim or claims.

Sec. 7. Every person shall have the privilege of the water to work his or their claim on any tributary or creek.

Sec. 8. Any person taking the water from its natural channel, shall, when required, be compelled to leave at least one sluice head of the water running therein.

Sec. 9. Thirty square inches of running water shall constitute a sluice head.

Sec.10. One hundred feet front and running back in a straight line to the summit of any hill shall constitute a tunnel or hill claim.

Sec. 11. Any person or company who may discover hill diggings, shall be entitled to one additional hill claim.

Sec. 12. Any citizen may hold one creek, one hill, and one ravine claim.

Sec. 13. When two or more claims join together and are worked by companies, such companies can work any portion of such claims as they may deem expedient, and not forfeit the rest.

Sec. 14. Any person disabled by sickness shall not forfeit his claim.

Sec. 15. Claims shall be distinctly marked by notices at each end of the bounds thereof, in the name of the claimant, and no more ground shall be deemed claimed than is included in such notices.

Sec. 16. Any person who wantonly shall haul down or destroy any notice on any mining claim or laws of this district, shall be fined in any sum that the miner's court of inquiry may direct.

Sec. 17. No person who may hold a mining claim shall forfeit said claim by reason of being compelled to leave said district for provisions for himself or company, provided he does not absent himself for a longer period than twenty-five days.

Sec. 18. Any person or persons who may have grievances to settle shall do so by arbitration, each party choosing a person to settle the matter, and in case they cannot agree, for the two arbiters to choose a third arbiter, a majority of whom shall settle the difference, and in case they do not settle the matter, then a call of a miner's meeting, who shall assemble and settle such matters as they may deem fit.

Sec. 19. There shall be a recorder of the mining district chosen, who shall be paid two dollars and fifty cents for recording each mining claim. The recorder shall have custody of the books and laws of this district and shall hold his office at the will of the majority of the district.

Sec. 20. Every person holding a mining claim shall have it recorded within thirty days from the date of the passage of these laws and from that time all claims shall be recorded when taken up.

Sec. 21. Every person's name holding a mining claim shall appear on the record.

Sec. 22. Any person having a claim shall have it represented according to law.

Sec. 23. No transfer of sale of claim shall be considered valid unless it appears on the record, and the recorder shall receive two dollars and fifty cents for the transfer.

Sec. 24. These laws to be amended hereafter at the option of the miners and to take effect from and after their passage.

M. Moore, Ch'mn
John W. Park, Sec'y

SILVER CREEK HOTEL RULES

In July, 1889, the Snohomish Weekly, The Eye, announced that the mountains were alive with prospectors and a number of gold-bearing lodes had been found west of Silver Creek. Frank Peabody and Joe Pearsall had an "adventure with a bear; neither bear nor men suffered a great deal from the meeting." Among the guests at the Silver Creek Hotel that week was F.W. Peabody. The hotel was probably located at Mineral City.

The following were the guest rules posted at the Silver Creek Hotel - "Which transient guests are required to observe faithfully:"

1. If the fleas are troublesome you'll find cloroform in a bottle on the shelf.
2. Gents goin' to bed with their boots on will be charged extra.
3. Three raps at the door means there is murder in the house, and you must get up.
4. Please rite your name on the wall paper so we know you've been here.
5. The other leg of the chair is in the closet if you need it.
6. If that hole where the pane of glass is out is too much for you, you'll find a pair of pants back of the door to stuff in it.
7. The shooting of a pistol is no cause for any alarm.
8. If you're too cold, put the oilcloth over your bed.
9. Karoseen lamps extra; candles free, but they musn't burn all night.
10. Don't tear off the wallpaper to lite your pipe with. Nuff of that already.
11. Guests will not take out them bricks in the mattress.
12. If it rains through that hole overhead, you'll find an umbreller under the bed.
13. Two men in a room must put up with one chair.
14. Please don't empty the sawdust out of the pillers.
15. Don't kick about the mice. We don't charge extra.
16. If there's no towell handy use a piece of the carpet.

THE GOLD PAN

The gold pan has served for centuries as a valuable tool for the recovery of gold, almost to the extent of being considered synonymous with prospecting. Used with skill, it is a reliable means of testing for the presence of gold. And it has the advantage of being light enough to conveniently carry into remote areas.

Modern gold pans are constructed in many shapes and sizes. Some are made of durable plastic and have riffle traps built into the side. The traps are helpful in saving heavier concentrates which contain fine gold. For those interested in gold panning, the following steps are recommended:

1. In gold producing areas, bedrock crevices provide natural traps for gold. A screwdriver or teaspoon can be used to dig out the compacted layer of dirt and gravel from between sections of rock. Be sure to rinse over the pan any loose pieces of bedrock. Quite often gold will be clinging to the clay or mud layer on the sides of the rock. The gathered material can then be panned out.

2. With the pan about half full, find a comfortable spot at the river's edge where you can easily extend the pan into the running water. Submerge the pan completely. Using your hand, stir through the pan and toss out any larger rocks. Break up any small lumps of clay or dirt which might have gold clinging to them.

Veteran prospector, Bob Lorz, demonstrates the use of the Garrett Gravity Trap gold pan.

3. Lift the pan slightly out of the water. Now with an oval-circular motion (do not swirl the pan), agitate the material in the pan allowing the gold to begin filtering downward. When you hear aggregate scraping against the bottom, this indicates the loose suspension of materials which allows the gold to travel downward.

4. As you continue to agitate, change to a side-to-side motion while at the same time tipping the forward side of the pan (the side away from you) downward, thus moving the discard material to the edge. As it reaches the edge, scoop in small portions of water and carefully float the top layer of sand and gravel from the pan.

5. When you have washed a safe amount from the top, scoop in a larger amount of water and return the material to a horizontal position in the pan. Begin the cycle all over again - agitating the material, tipping the pan forward with a side-to-side motion, floating off the top layer, and returning to a horizontal position. Each time you return to a horizontal position, with your hand, scoop off any larger pebbles or rock.

6. As the bottom of the pan is exposed, black sand should begin to appear. If it does, its a good sign. Pan off everything except the black sand, then swirl a small amount of water in the pan. If any gold is present, it will be revealed as the black sand moves along. Develop the skill of panning off the black sand without losing the gold. Also, don't be fooled by fool's gold (iron pyrite) which will appear as small pale-yellow chunks in the pan. Remember that these have angular edges and are brittle.

7. When the whole process is complete, all you should have left is the gold with a small residue of black sand. You have now achieved an important step in the broad endeavor of prospecting - how to use the gold pan.

WEIGHING YOUR GOLD

The last step in the recovery of gold is cleaning and weighing it. When the gold is dry, a magnet (shielded with paper) can be used to remove the remaining magnetite particles. Sometimes fine gold is picked up with other particles, so be careful to retrieve it. After the gold is separated, a balance scale or reloading powder scale can be used to weigh it. Balance scales are available with Troy counterweights for weighing precious metals. The Troy system is based on the weight of 1 pound equal to 12 ounces, 1 ounce equal to 20 pennyweight, and 1 pennyweight equal to 24 grains.

Probably the most popular use of native gold is jewelry. Nuggets can be crafted into strikingly beautiful rings and pendants. Given as a symbol of affection, natural gold jewelry is a novelty always in demand. Whether refined or unrefined, used as a medium of exchange, as a symbol of affection, or as a protective crown over a broken tooth, gold will always be considered a valuable commodity. And the challenge of finding raw gold, even small traces, can be a rewarding experience for the young and old alike.

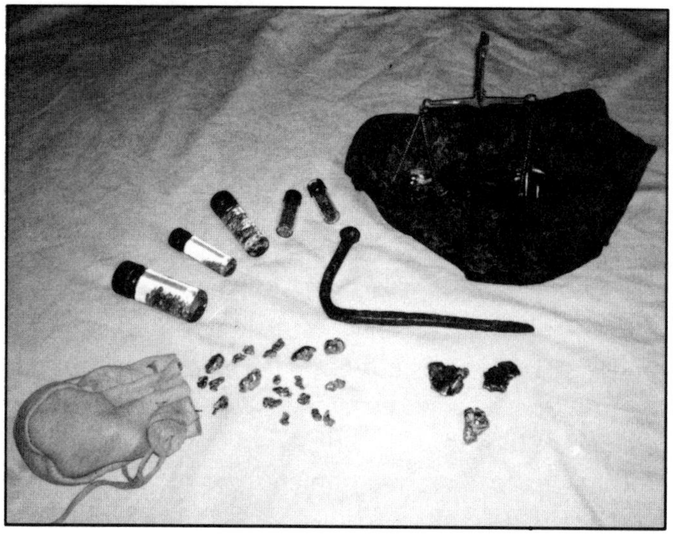

Gold scales with nuggets and vials of gold. The hand-wrought nail was bent to form a crevicing tool. I found the nail stuck in bedrock at the bottom of the Similkameen River. It may have been there since the 1800's.

REGULATIONS

Regulations governing mining and prospecting activities are continually being reviewed and updated, and these vary from state to state. It is therefore a good idea for anyone interested in dredging or other prospecting activities to become informed as to what he or she may or may not do within a specific area. In Washington State, the Department of Fish and Wildlife has determined by classification the types of activities permitted on a given stream and the time of year one is allowed to work. This is spelled out in the small booklet, "Gold and Fish." If a person wishes to prospect a stream or river, a portion of which may be under lease or contract from the state, he or she should determine its status and make proper arrangements before prospecting.

In addition to state owned streams and rivers, there are many gold producing streams which are located on private land. One should always exercise the courtesy of acquiring permission before prospecting these streams or entering private property.

State and private lands are not open to the location of mining claims. Federal lands which are open to mining claims are designated "open public land." There are a number of areas which are <u>not</u> open or have been withdrawn. Some of these are federal wildlife refuges, national parks, national recreation areas, wilderness areas, and Indian Reservations. It is a good idea to inquire at the nearest ranger station before entering the national forests.

The authors of this book will not assume responsibility for anyone's actions as related to prospecting or mining activities.

BETTER THAN GOLD

Better than gold is a conscience clear,
Though toiling for bread in an humble sphere
Doubly blest with content and health
Untied by the lust and cares of wealth,
Lowly living and lofty thought,
Adorn and ennoble a poor man's cot -
For man and morals, on nature's plan
Are the genuine test of a gentleman.

Better than gold is the sweet repose,
Of the sons of toil, when their labors close,
Better than gold is the poor man's sleep,
And the balm that drops on his slumbers deep,
Brings sleepy draughts to the downy bed,
Where luxury pillows his aching head;
His simpler opiate labor deems
A shorter road to the land of dreams.

Better than gold is the thinking mind
That in the realm of books can find
A treasure surpassing Australian ore,
And live with the great and good of yore,
A heart that can feel for a neighbors woe,
And share his joys with a genial glow;
With sympathies large enough to enfold
All men as brothers - is better than gold

Better than gold is a peaceful home,
Where all the fireside charities come
The shrine of love and the heaven of life,
Hallowed by mother, or sister, or wife,
However humble the soul may be,
Or tried by sorrow with heaven's decree,
The blessings that never were bought or sold,
And center there, are better than gold.
- from the Northern Star, Jan. 20, 1877.

Psalm 19: 9b-10a, 1000 B.C.

The Ordinances of the Lord are true,
and righteous altogether.
More to be desired are they than gold,
even much fine gold. (RSV)

BIBLIOGRAPHY

Part 1

1. Ex. Doc. No. 56. House of Rep. Reports of Explorations and Surveys to Ascertain the Most Practical and Economical Route for a Railroad from the Mississippi River to the Pacific Ocean. According to Act of Congress of Mar. 3, 1853, May 31, 1854, and Aug. 5, 1854. Vol. XII, Book 1. Washington: Thomas H. Ford, 1860. P. 140.

2. Ex. Doc. No. 56. P. 257.

3. From the American Railroad Journal, July 31, 1858. "Northern Pacific Railroad." The Pioneer and Democrat, Olympia: Washington Territory, Friday, 1 Oct. 1858.

4. Trimble, William J. The Mining Advance Into the Inland Empire. Diss. U. of Wisconsin 1914. Madison, Wisconsin, 1914. p. 16.

5. "Letter From Frazier River." 8 Aug. 1858. The Pioneer and Democrat, Friday, 20 Aug. 1858.

6. "The Swauk Mines." The Northern Star. Snohomish City, W.T. Saturday, 17 Aug. 1878.

7. "Great Excitement." Immense Fields of Gold Discovered on all the Rivers West of the Cascade Mountains. Pioneer and Democrat, Friday, 2 July, 1858.

8. Morse, E. "The Mineral Resources of Snohomish County." The Eye. Saturday, 20 April, 1889.

9. Steele, Richard F. History of North Washington: An Illustrated History of Stevens, Ferry, Okanogan, and Chelan Counties. State of Washington: Western Historical Publishing Company, 1904. pp. 38-39.

10. Trimble, p. 23.

11. "From the Colville Region." The Pioneer and Democrat. Friday, 3 Sept. 1858.

12. "Rich Placer Bars." The Everett News. Saturday, 14 May, 1892.

Part II

1. Louis, Loretta. "Ruby City: The Life and Death of a Mining Town." Okanogan County Heritage." June, 1963. pp. 3-16.

2. Bethune, George A. Mines and Minerals of Washington. Olympia: O. C. White, State Printer. 1891. p. 52.

3. Fischer, Arthur Homer. A Summary of Mining in the State of Washington. Bulletin No. 4, Seattle: Published by the University. Nov. 1918. p. 27.

4. Steele, Richard F. History of North Washington: An Illustrated History of Stevens, Ferry, Okanogan, and Chelan Counties. State of Washington: Western Historical Publishing Company, 1904. p. 540.

5. "An Expensive Product, Silver Costs More than a Dollar an Ounce." Snohomish Eye, Friday, 3 March, 1893.

6. Bethune. p. 6.

7. Steele. p. 513.

8. Townsend, W. H. "The Mines Rapidly Developing - Rich Strikes Recently Made." Spokane Review. 1, Jan. 1892.

9. Hatchet, M.D., et al. Told by the Pioneers. Vol. 11 Printed under W.P.A. Sponsored Federal Project No. 5841. 1938. p. 198.

10. Briley, Ann. "Hiram F. Smith, First Settler of Okanogan County." Pacific Northwest Quarterly, 43, No. 3 (1952), p. 227.

11. Meany, Edmond S. History of the State of Washington. New York: The McMillan Company. 1910. p. 153.

12. Steele. p. 486.

13. "Death of H.F. Smith." The Pioneer of the Okanogan Country Passes Away. The Seattle Post-Intelligencer, Sunday, 10 Sept. 1893.

14. Umpleby, Joseph B. Washington Geological Survey, Bulletin No. 6. Olympia: E. L. Boardman, Public Printer. 1911. p. 60.

15. "Rich Gold Discoveries: One Dollar to the Pan." Pioneer and Democrat, Friday, 14 Sept. 1860.

16. "Placer and Quartz: The Swauk Placer Mines." Spokane Review, 20 Dec. 1891.

17. Weaver, Charles E. Geology and Ore Deposits of the Blewett Mining District. Bulletin No. 6. Olympia: E.L. Boardman, Public Printer. 1911. p. 71.

18. Weaver. pp. 69-70.

19. Hoover, Herbert Clark., et. al. Georgius Agricola DE RE METALLICA, translated from the First Latin Edition of 1556. New York: Dover Publications, Inc. 1950. p. 282.

20. Photo reprinted from: Moen, W. S. 1982, "The Mineral Industry of Washington - Highlights of its Development," 1853-1980: Washington Division of Geology and Earth Resources Information Circular 74, p. 7.

21. Weaver. pp. 68-69.

22. Steele. p. 515.

23. "Rich Cascade Mines." The Everett News, Friday, 25 Dec. 1891.

24. Letterpress copybooks. Manuscripts and University Archives, Allen Library, University of Washington.

25. Letterpress copybooks, U.W.

26. Fischer, p. 28.

27. Map reprinted from: Moen, W. S., 1969, "Mines and Mineral Deposits of Whatcom County, Washington" Washington Division of Mines and Geology. Bulletin 57, p. 96.

28. Moen, W. S., 1969, Bulletin 57, front cover. Original from the Washington Department of Commerce and Development.

29. Blonk, Hu. "Early-Day Harts Pass Mine Camp Tough Spot." The Wenatchee Daily World, Sunday, 10 Apr. 1960.

30. Olling, Don. "Touchy Job In Powerhouse." Okanogan County Heritage. Vol. 13, No. 1. 1974-75. pp. 15-16.

31. "Charles Ballard Was Most Active Useful Citizen." Auburn News. 3 Jan. 1935.

32. Ballard, Charles H. Pioneer Experiences of the Ballard Family on White River. King County, Washington. White River Valley Historical Society. 1914-15, p. 5.

33. Ballard, p. 5.

34. Ballard, p. 17.

35. Steele, p. 557.

36. Diana Hottell, "The Rise and Fall of the Gold Hill Mine:" The Okanogan Heritage. Winter, 1984-1985.

37. Bagley, Clarence B. History of Seattle - From the Earliest Settlement to the Present Time. Chicago: The S. J. Clarke Publishing Co. Vol. 11. 1916. pp. 692-93.

38. "Ballard is Now Part of Seattle." The Seattle Post Intelligencer. 30 May, 1907.

39. Information provided by the Hecla Mining Company. Coeur d' Alene, Idaho.

Bibliography
Part III

1. Bethune, George, A. Mines and Minerals of Washington. Olympia: O.C. White, State Printer. 1891. p. 6.

2. Snowden, Clinton A. History of Washington, The Rise and Progress of an American State. Vol. 4. New York: The Century History Company, 1909. pp. 39.

3. Meeker, Ezra. Ventures and Adventures of Ezra Meeker. Seattle: Rainier Publishing Company. 1909. pp. 245-246.

4. Meeker. p. 247.

5. "The Skagit River and Country." Pioneer and Democrat. Friday, 20 Aug. 1858.

6. "The Snoqualmie Route to the Mines." Pioneer and Democrat. Friday, 13 Aug. 1858.

7. "Latest News From the Mines." Pioneer and Democrat. Friday, 20 Aug. 1858.

8. "Early History of Snohomish River and Vicinity Recalled by E. C. Ferguson." The Everett Daily Herald. Tuesday, 14 Jan. 1936.

9. Williams, John H. The Canoe and The Saddle or Klalam and Klickatat by Theodore Winthrop. Tacoma: John H. Williams, 1913. p. 282.

10. "From the Mines: America City, Nov. 26, 1860." The Washington Standard. Olympia, Washington Territory, 5 Jan. 1861.

11. Hodges, L.K. Mining in the Pacific Northwest. Seattle: The Post-Intelligencer, 1897. p. 67.

12. "Placer and Quartz: The Swauk Placer Mines." Spokane Review, 20 Dec. 1891.

13. Hodges. p. 67.

14. "Placer and Quartz: The Swauk Placer Mines." Spokane Review, 20 Dec. 1891.

15. "The Swauk Mines," The Northern Star, Saturday, 17 Aug. 1878.

16. Wilson, Eugene B. Hydraulic and Placer Mining. New York: John Wiley and Sons. 1900. pp. 104-108.

17. State of Washington, Department of Natural Resources. Aquatic Land Management Plan for the Okanogan- Rivers. Olympia, Washington. 1982. p. 31.

18. "Mines and Mining: Tricks of the Trade Practiced on Investors" The Eye, Thursday, 3 Jan. 1896.

Author's Note:

The history of Okanogan County, including the subject of mining, is well presented in the book Late Frontier. A History of Okanogan County, Washington by Bruce A. Wilson, published by the Okanogan County Historical Society, 1990.

INDEX

H

I

J

K

L

M

N

O

P

Q

R

S

T

U

V

W

Y